CHEMISTRY MATTERS

CONTRIBUTORS

MYRNA MITCHELL

JIM WILSON

RON KOOP

TRACEY LIEDER

ORIGINAL WRITING TEAM

BETTY CORDEL

DAVID MITCHELL

SHARON BOWERS

EVELYN KOVALICH

JOHANN WEBER

MAUREEN WATTS

LINDA LIEBLER

RENEE MASON

...and many others

CHEMISTRY MATTERS

This book contains materials developed by the
AIMS Education Foundation. **AIMS**
(**A**ctivities **I**ntegrating **M**athematics and **S**cience)
began in 1981 with a grant from the National Science
Foundation. The non-profit AIMS Education Foundation publishes hands-on
instructional materials (books and the monthly magazine) that integrate curricular
disciplines such as mathematics, science, language arts, and social studies. The
Foundation sponsors a national program of professional development through which
educators may gain both an understanding of the AIMS philosophy and expertise in
teaching by integrated, hands-on methods.

ISBN **1-932093-03-6**
Printed in the United States of America

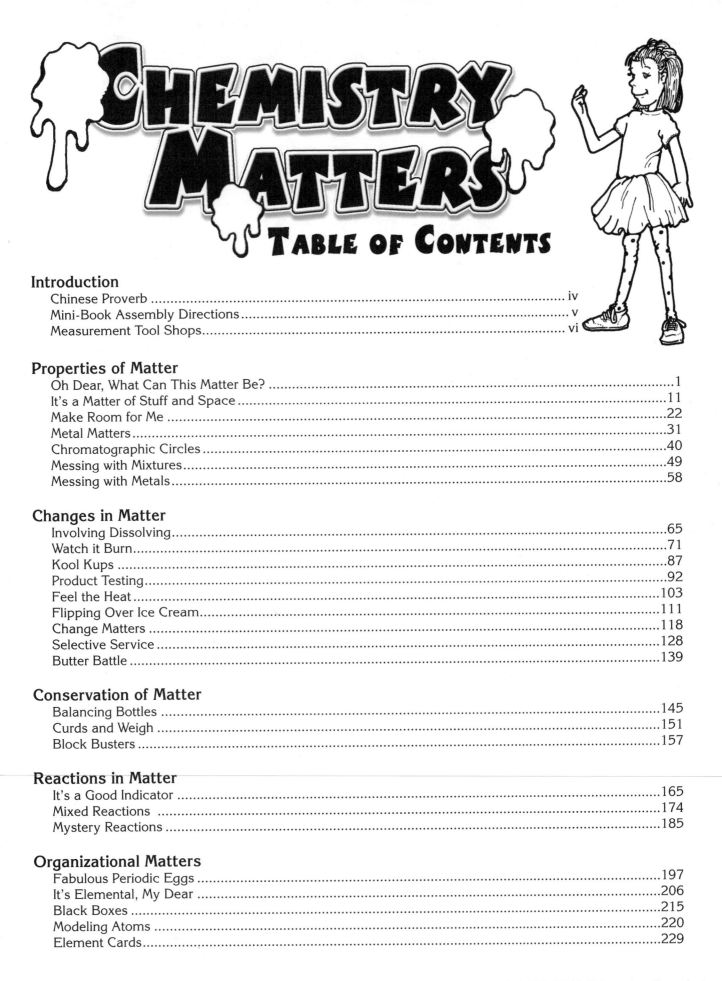

CHEMISTRY MATTERS

TABLE OF CONTENTS

I Hear and
I Forget,

I See and I
Remember,

I Do and I
Understand.

Chinese Proverb

Mini-Book Assembly Directions

Measurement Tool Shop
Linear

How far? How long? How many? How much? We constantly deal with questions like these in our daily lives. Measurement gives us a way to communicate clearly about things around us. The United States uses both customary units and metric units of measure. The focus of this **Tool Shop** is to learn how to better use metric linear units. The metric system is a base-ten system, the same numbering system we use to count. We will be using millimeter (mm) and centimeter (cm) units.

Linear Measurement

Linear measurement can tell us about many attributes of matter. We can measure the attributes of length, circumference, and perimeter using linear units. This is a centimeter. |1 cm| It is about the same distance across as a large paper clip. A centimeter can be divided into smaller units. This is a centimeter that has been divided into ten equal units. The name of this unit is called a millimeter.

10 mm

Linear Metric Tool Shop

Materials needed: a small paper clip, a large paper clip, a can, a small ball, a piece of string, and two blocks of wood

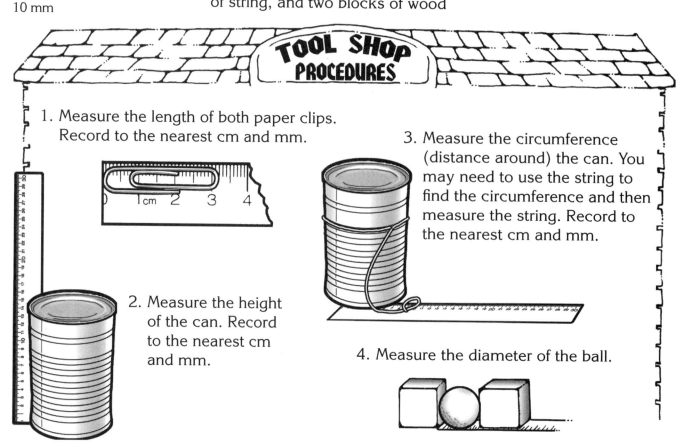

TOOL SHOP PROCEDURES

1. Measure the length of both paper clips. Record to the nearest cm and mm.

2. Measure the height of the can. Record to the nearest cm and mm.

3. Measure the circumference (distance around) the can. You may need to use the string to find the circumference and then measure the string. Record to the nearest cm and mm.

4. Measure the diameter of the ball.

Measurement Tool Shop
Mass

How far? How long? How many? How much? We constantly deal with questions like this in our daily lives. Measurement gives us a way to communicate clearly about things around us. The United States uses both customary units and metric units of measure. The focus of this **Tool Shop** is to learn how to determine the mass of objects using a balance and masses. The mass of objects is most commonly recorded in milligrams (mg), grams (g), or kilograms (kg). These are the metric units of mass. The metric system is a base-ten system, the same numbering system we use to count.

Mass Measurement

Mass is the amount of matter or "stuff" in something. No matter where in the universe something is, or what it is doing (moving or staying still), its mass will not change. Mass is also a measure of the inertia (a resistance to a change in motion) of something. The more mass, the harder to change its motion (stop, start, or turn). This is why it takes a longer distance to stop a car than it does to stop your bike. Inertia is also why trains and trucks take a while to get moving from a stop.

Mass Metric Tool Shop

Materials needed: a balance, a set of masses, one wooden block, and four classroom objects

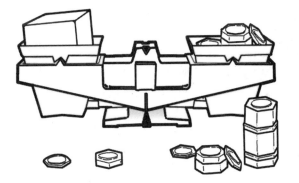

Tool Shop Procedures

1. Place the wooden block on one of the pans in the balance. Place masses on the other pan until the needle of the balance lines up. Record the mass of the block in grams.

2. Predict the order of the four items from the lightest to the heaviest. Find the mass of each and record the mass in grams.

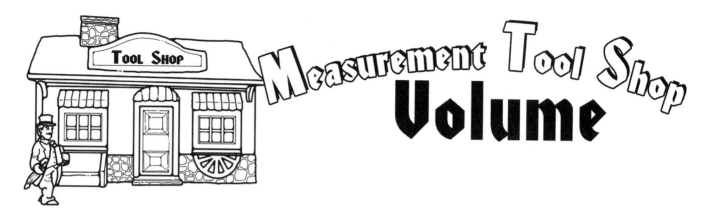

Measurement Tool Shop
Volume

How far? How long? How many? How much? We constantly deal with questions like this in our daily lives. Measurement gives us a way to communicate clearly about objects and events. The United States uses both customary units and metric units of measure. The focus of this **Tool Shop** is to learn how to determine the volume of regular- and irregular-shaped solids as well as the volume of liquids. Volume measurement is recorded in cubic units such as cubic centimeters (cm^3) or milliliters (mLs). These are units in the metric system. The metric system is a base-ten system, the same numbering system we use to count.

Volume Measurement

Volume measurement tells us how much space an object takes up. We use graduated cylinders and beakers to find liquid volume as well as capacity.

There are two methods we can use to measure solid volume. You can measure regular-shaped objects like boxes and balls and use a formula to calculate volume. The other way to find volume of objects is through displacement.

The principle behind displacement is that two objects cannot occupy the same space at the same time. To use displacement to find the volume of an object, you need a graduated cylinder or beaker partially filled with water. The level of the water in the container is read and recorded. The object you want to find the volume of is placed (submerged) in the container. The water level will rise as the object is placed in the container. The new water level is read and recorded. The difference between the two levels is the object's volume. Graduated cylinders and beakers are often scaled in milliliter units. In the metric system 1 mL of water has a cubic volume of 1 cm^3.

Measurement Tool Shop Volume

Volume Metric Tool Shop

Materials needed:
two boxes or blocks of wood, 100 mL graduated cylinder, 250 mL beaker, a golf ball, rocks, two cups of liquids, and one empty cup.

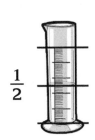

$\frac{1}{2}$

Tool Shop Procedures

1. Fill a graduated cylinder half full of water. Look at the side of the graduated cylinder. Note the scale and read and record the amount of water. Find the volume in the two cups as well as the capacity of the empty cup.

read and record

read and record water level left

or

full cup

read and record

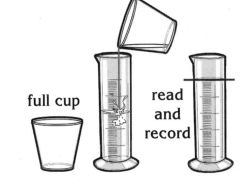

The difference in the two levels is the volume of liquid in the cup.

2. The formula for solid volume for a box-like object is length x width x height = volume. The geometric term for this type of object is a rectangular solid. Find the volume of the two boxes. Record the measurements to the nearest centimeter, then calculate to find the volume in cubic centimeters (cm³).

3. To find the volume of the rock, you need to use the process of displacement. Pour 150 mL of water into the 250 mL beaker. Place the rock in the beaker of water. The water level will rise and then you read the new water level. The difference between 150 mL and the new level is the volume of the rock. You can also use the graduated cylinder to find the volume of the object if it will fit in the cylinder. Find the volume of the two rocks. Now find the volume of the golf ball. Use the tip of your pencil to push the golf ball under the surface of the water. Get as little of the pencil tip in the water as possible.

read and record

Temperature

How far? How long? How many? How much? We constantly deal with questions like this in our daily lives. Measurement gives us a way to communicate clearly about objects and events. The United States uses both customary units and metric units of measure. The focus of this **Tool Shop** is to learn how to determine the temperature of objects and experiments using a Celsius thermometer. Celsius is the temperature scale used in the metric system. The metric system is a base-ten system, the same numbering system we use to count.

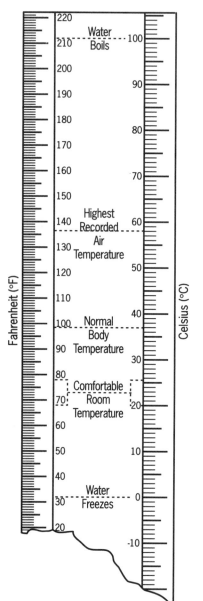

Temperature Measurement

The United States uses both Celsius (C) and Fahrenheit (F) when recording temperature. You are probably most familiar with Fahrenheit. Examine the drawing on the left. This will help give you some benchmarks for Celsius.

The symbol ° means degree. The temperature 30°C should be read as 30 degrees Celsius. One of the most important things to remember when using a thermometer is to first check the scale. This means you must check to see how the thermometer is calibrated or divided. The two drawings below show two different scales.

They are calibrated or divided into 1-degree intervals and 5-degree intervals. Pick up the thermometer you are going to use in this tool shop. Notice that on the thermometer not all the lines have numbers on them. You need to know the scale so you can determine what each line equals. The final thing to keep in mind in using a thermometer is to allow time for the thermometer to adjust to the temperature of the object or experiment. Two minutes is usually enough time.

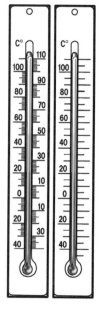

Temperature

Temperature Metric Tool Shop

Materials needed: thermometer, cup, spoon, water, and ice

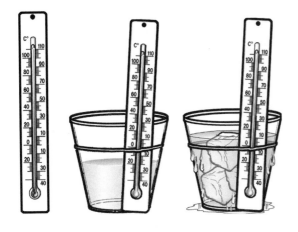

Tool Shop Procedures

1. Take the temperature of the air. Allow the thermometer time to adjust to the temperature by waiting two minutes before you read it.
2. Fill the cup half full with water. Find the temperature of the water.
3. Place some ice in the cup and record the temperature every two minutes. You will read the temperature every two minutes until ten minutes have passed.

Air temperature

Temperature of the water in the cup

Temperature after adding ice

After 2 minutes

After 4 minutes

After 6 minutes

After 8 minutes

After 10 minutes

Oh Dear What Can This Matter Be?

Topic
Physical Properties of Matter

Key Question
What can you learn about objects by classifying them?

Learning Goals
Students will:
1. classify objects based on physical properties,
2. identify the three states (phases) of matter, and
3. use a Venn diagram to classify objects based on their states (phases) of matter.

Guiding Documents
Project 2061 Benchmarks
- *Objects can be described in terms of the materials they are made of (clay, cloth, paper, etc.) and their physical properties (color, size, shape, weight, texture, flexibility, etc.).*
- *Atoms and molecules are perpetually in motion. Increased temperature means greater average energy of motion, so most substances expand when heated. In solids, the atoms are closely locked in position and can only vibrate. In liquids, the atoms or molecules have higher energy of motion, are more loosely connected, and can slide past one another; some molecules may get enough energy to escape into a gas. In gases, the atoms or molecules have still more energy of motion and are free of one another except during occasional collisions.*

NRC Standards
- *Objects have many observable properties, including size, weight, shape, color, temperature, and the ability to react with other substances. Those properties can be measured using tools, such as rulers, balances, and thermometers.*
- *Objects are made of one or more materials, such as paper, wood, and metal. Objects can be described by the properties of the materials from which they are made, and those properties can be used to separate or sort a group of objects or materials.*
- *Materials can exist in different states—solid, liquid, and gas. Some common materials, such as water, can be changed from one state to another by heating or cooling.*

NCTM Standards 2000*
- *Collect data using observations, surveys, and experiments*
- *Represent data using tables and graphs such as line plots, bar graphs, and line graphs*

Math
Data analysis
    Venn diagram

Science
Physical science
    matter

Integrated Processes
Observing
Comparing and contrasting
Classifying
Communicating
Predicting
Collecting and recording data
Organizing data

Materials
For each student:
    1 lunch-size paper bag

For each group:
    examples of solids (see *Management 1*)
    examples of liquids (see *Management 2*)
    examples of gases (see *Management 3*)
    2 balloons, any size or shape
    Grouping Circles (see *Management 5*)
    sheets of chart paper or newsprint
    needle with large eye
    crochet thread
    Marvelous Matter mini-book (see *Management 7*)

Background Information
    Matter is defined as anything that has mass and takes up space. This definition involves two properties. The first is mass, which is the amount of material in an object and is measured with a balance. The second is volume, which is the amount of space taken up by an object. Matter has physical properties that help distinguish one kind of matter from another kind. Some properties of matter are color, shape, composition, size, density, as well as others.

Phase is another important physical property of matter. Matter on Earth most commonly exists in three phases: solid, liquid, and gas. Solids have a definite shape and a definite volume. Liquids have a definite volume but take the shape of the container they are in. Gases have no definite shape or volume. The spacing between the molecules of a substance determines which phase that substance is in. Solids are the most tightly packed followed by liquids and gases. A change in the spacing between the molecules results in change of phase. Heat and pressure can cause a change in the spacing.

In the first part of this activity, the students will be engaged in utilizing the scientific process skill of classification. There is no one correct classification scheme. Students' ideas should determine classification. It is possible to classify the objects in different categories each time a new classification scheme is devised. The important idea is that scientific thinking involves determining and using categories. The second part has students examining the three phases of matter using a Venn diagram.

Management
1. Tell the students they will need to bring in a collection of five objects that will fit inside a brown lunch bag. Encourage the students to bring in a wide variety of items.
2. Prepare to have some examples of liquids for the students to use during *Part Two* of the activity. Some common school-related examples are glue, paint, and milk. Bottles of water as well as fruits that contain juice (apples and oranges) are other items that could be used. Have at least three examples of liquids for each group.
3. You may want to bring in a helium-filled balloon for the students to use in the classification for *Part Two*.
4. The inflated balloon provides an example of a gas, but point out that the space above the water in a water bottle is also an example of a gas.
5. Grouping circles may be purchased from AIMS. They can also be constructed by using three different colors of yarn. Cut the yarn into one meter lengths, then tie the ends to form a circle.
6. Students should be placed in groups of four or five.
7. The student mini-books are constructed by first folding the four pages in half along the 11-inch dimension so that the print is on the outside of the fold. The pages are then stacked in order with the book's cover on the bottom and the center spread page on top (pages 6 and 7). Flip the book pages over. On the cover locate the four holes along the center spine. These are to be used as a guide to sew the book together. Fold the pages in half after you have bound them together. Instead of sewing the spine, you can use a rubber band to hold the spine together.

Procedure
Part One
1. Ask the *Key Question* and state the first *Learning Goal*.
2. Distribute chart paper to each group.
3. Tell each person to share with the group the collection of objects that they brought. Have the group develop three different ways to classify the objects. For example, the students could use color, size, composition, etc. For each classification, have students record the categories they created and list the objects that fit into each category on the chart paper.
4. Have the groups share their classification systems and their reasoning for their categories and the objects classified.
5. Explain that categories used in classification are often called properties and that all the objects in each collection are matter.
6. Have the students read the mini-book on matter. After the students read the book, discuss physical properties.
7. Explain that the categories that the groups probably used are called physical properties (color, shape, composition, size, texture). List the categories students choose under the title: physical properties.
8. Each group should choose a physical property from the list that the class generated and show how the objects their group has can be classified.
9. Ask the students for another way in which objects could be classified based on what they read in the mini-book. (Objects on Earth are typically classified as a solid, liquid, or gas.)

Part Two
1. Ask the *Key Question* and state the second and third *Learning Goals*.
2. Have the students select five objects from the ones the group brought in and then ask them to put the others away. Distribute the liquids you have prepared to each of the groups.
3. Give each group two of the grouping circles to make a two-circle Venn diagram. Ask students to classify the objects into solid and liquids. If the objects show both phases, such as a carton of milk, they would be placed in the intersections of the circles.

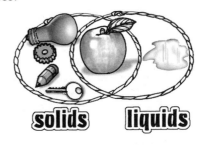

4. When the students have completed classifying the materials, distribute the third grouping circle and tell them it will be used to classify the third phase of matter, gas. Arrange the grouping circles so that the three circles are arranged as illustrated.

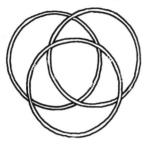

5. Give each group the two balloons. Tell the students to inflate one and leave the other as it is. Ask them to now classify the two balloons. Ask them if they will now need to move some of the other objects since you have added the third hoop. (The bottles of glue and milk cartons have air in the containers. Make sure the students understand that a gas is present in these examples.)

6. When the students have classified all the objects, have them record on the student Venn sheet the name of the objects they classified in the correct region of the Venn diagram.

7. Ask the students to complete a phase scavenger hunt. The goal is to have at least one example in all regions of the Venn diagram. Tell them to record the examples in words and sketches.

Connecting Learning

1. What are physical properties of matter?
2. What are the three most common states (phases) of matter on Earth? [solid, liquid, and gas]
3. Why do you think it was easier to find solids than liquids and gases? [Liquids and gases both need containers to hold them.]
4. How did the balloon help in this activity? [It served as a container to hold a gas.]
5. How could you convince someone that all three states (phases) of matter are present in a can of soda? [The can is made of metal, which is a solid; the soda is a liquid; and there is air in the can as well as gas bubbles in the soda.]

Evidence of Learning

1. Watch as students physically classify the objects while using the grouping circles. Look for correct placement as well as reasoning why objects were placed where they were.
2. Look at completed scavenger hunt sheet for accuracy in placement of matter based on phase.

* Reprinted with permission from *Principles and Standards for School Mathematics*, 2000 by the National Council of Teachers of Mathematics. All rights reserved.

Oh Dear What Can This Matter Be?

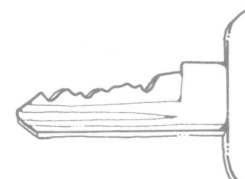

• Key Question •

What can you learn about objects by classifying them?

Learning Goals

Students will:

1. classify objects based on physical properties,
2. identify the three states (phases) of matter, and
3. use a Venn diagram to classify objects based on their states (phases) of matter.

solids   liquids

From the stars in the night sky to the sands on the beaches, from the air we breathe to the fizz in our sodas, from the planets in the solar system to the rocks and water on the Earth, everything in the entire universe is made of matter.

Matter in the gaseous state has neither a definite volume nor a definite shape. Gases expand to fill any container. Take a deep breath and hold it for a few seconds. You have just inhaled matter. Air is a gas and your lungs are the containers holding it.

Different kinds of matter have two things in common: they all take up space, and they all have mass.

Matter in the liquid state has a definite volume, but no definite shape. It assumes the shape of the container in which it is placed. If you were to have a glass of water on your desk right now, you would have two phases of matter in front of you, a solid glass and the liquid water.

2

11

4

Weight is actually the measure of the pull of Earth's gravity on the mass of an object. Mass can most easily be thought of as the amount of "stuff" in an object.

3

We can think of taking up space as you being seated in your chair at school. If you are sitting in your chair, no one else could take up that space but you. The amount of matter in any object is called mass. Mass is not quite the same thing as an object's weight.

6

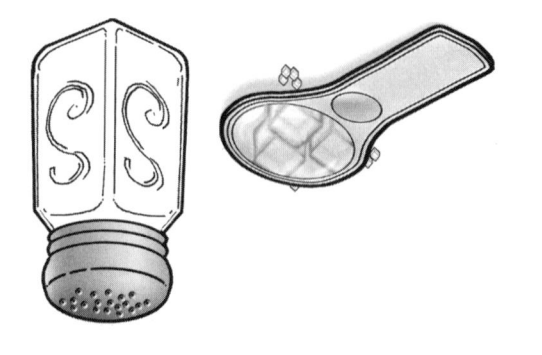

Intensive properties describe properties of matter that do not change. If you were to observe salt under a microscope or hand lens, you would notice it looks like small cubes. Salt has a crystalline structure, an example of an intensive property.

10

Matter commonly exists in one of three physical states or phases on the Earth. These states or phases are solid, liquid, and gas. Matter in the solid state has both a definite volume and a definite shape. A rock would be an example of a solid.

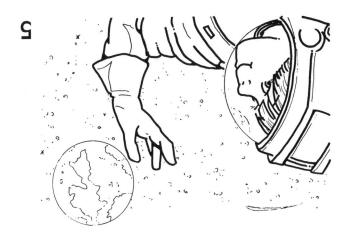

If you could go to the moon and weigh yourself, you would weigh less, but your mass would be the same. There would not be any less of you. We use a balance to measure mass and a spring scale to measure weight.

Matter can also be described based on its extensive and intensive prop- erties. Extensive properties are measurable. If you were finding the weight of an object, you would be finding an extensive property.

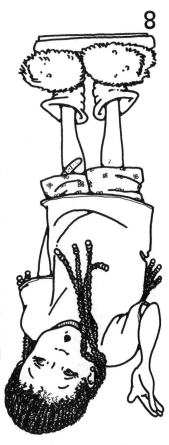

Matter is often described by its properties. Physical properties are used to describe the physical qualities of matter such as color, smell, taste, texture, and density. Chemical properties of matter describe the ability of a substance to be in a chemical reaction and to form a new substance. The rusting of metal would be an example of a chemical reaction.

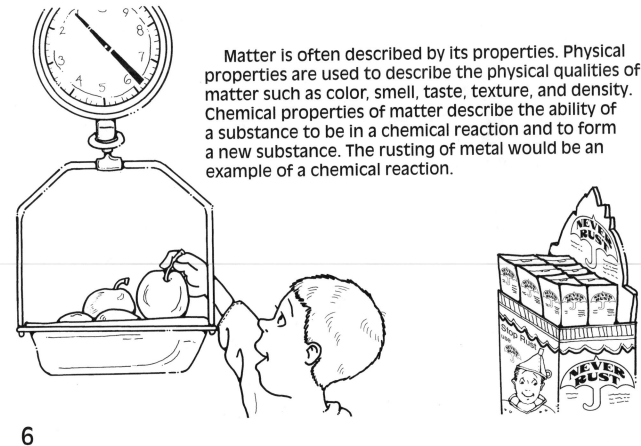

6

7

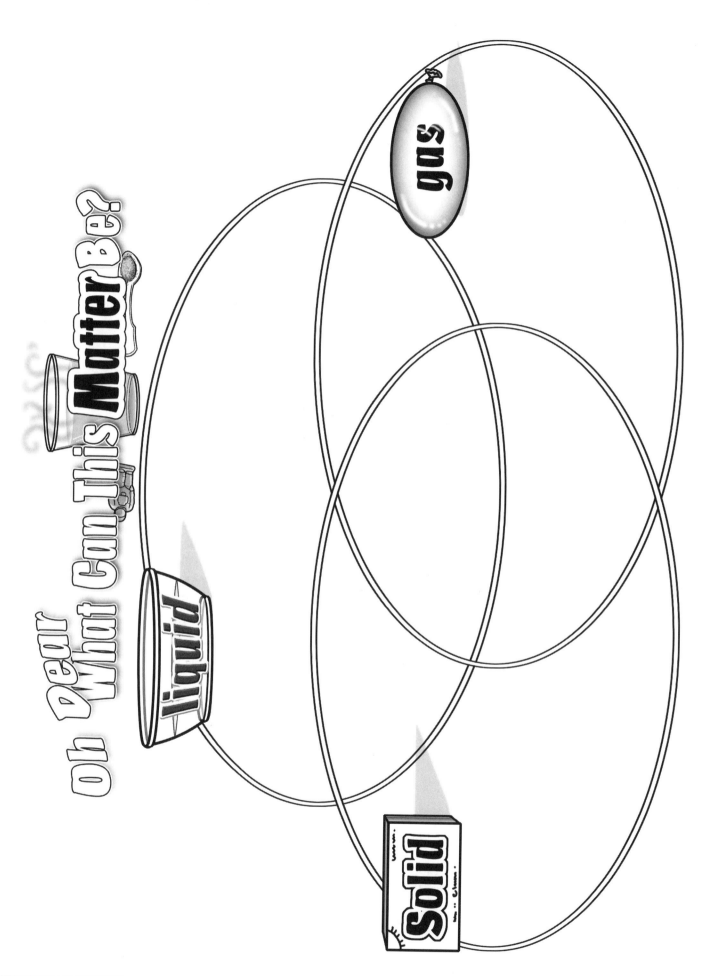

Oh Dear What Can This Matter Be?

Connecting Learning

1. What are physical properties of matter?

2. What are the three most common states (phases) of matter on Earth?

3. Why do you think it was easier to find solids than liquids and gases?

4. How did the balloon help in this activity?

5. How could you convince someone that all three states (phases) of matter are present in a can of soda?

6. What are you wondering now?

Topic
Defining matter

Key Question
What is matter?

Learning Goals
Students will:
1. define matter as anything that has mass and takes up space, and
2. identify examples of matter found in their classroom

Guiding Documents
Project 2061 Benchmark
- *Scientific investigations may take many different forms, including observing what things are like or what is happening somewhere, collecting specimens for analysis, and doing experiments. Investigations can focus on physical, biological, and social questions.*

NRC Standards
- *Objects have many observable properties, including size, weight, shape, color, temperature, and the ability to react with other substances. Those properties can be measured using tools, such as rulers, balances, and thermometers.*
- *Materials can exist in different states—solid, liquid, and gas. Some common materials, such as water, can be changed from one state to another by heating or cooling.*

NCTM Standards 2000*
- *Understand such attributes as length, area, weight, volume, and size of angle and select the appropriate type of unit for measuring each attribute*
- *Understand the need for measuring with standard units and become familiar with standard units in the customary and metric systems*
- *Recognize geometric ideas and relationships and apply them to other disciplines and to problems that arise in the classroom or in everyday life*

Math
Measurement
   volume

Science
Physical science
   matter

Integrated Processes
Observing
Comparing and contrasting
Classifying
Relating
Communicating
Generalizing

Materials
For the teacher:
   one 12-inch round balloon
   overhead transparency on the properties of matter

For each group of students:
   2 blocks of wood (see *Management 2*)
   1 ruler
   1 golf ball
   1 graduated cylinder or beaker
   1 9-ounce clear plastic cup
   1 balance
   1 set of masses
   2 12-inch round balloons

Background Information
   The word matter comes from the Latin word *mater*, which means "mother." Since everything is made from matter, matter could be called the mother of all things. Matter is defined as anything that has mass and takes up space. The definition defines matter in terms of two properties, mass and volume. Mass describes matter by the amount of material in an object. Mass is measured with a balance in grams. Volume is the second property of the definition of matter. It is described by the amount of space taken up by an object and is calculated through measurement or through the process of displacement.

Management
  1. The metric *Measurement Tool Shop* on *Linear, Volume* and *Mass* should be reviewed/taught

before completing this activity. This activity assumes that students can find the volume and mass of objects.

2. Wooden blocks can be cut from two-by-fours.

Procedure

1. Ask the students the *Key Question* and state the *Learning Goals*.

2. Distribute the wooden blocks, ruler, balances, and masses to each group. Tell students to make and record as many observations as possible of the block of wood, including its volume and mass. Tell the students to record observations on the student sheet.

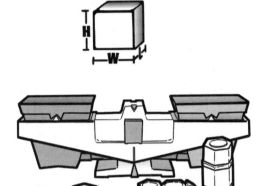

3. Distribute the golf balls, graduated cylinder, and cup. Tell the students to list all the characteristics they can about the golf ball. Direct the students to find the mass and volume measurements of the golf ball. Review or teach the students how to find the volume of the golf ball using displacement. Tell the students to compare the block of wood to the golf ball. Tell the students to record all observations on the student sheet. Ask, "How are the block of wood and the golf ball the same and how are the two different?"

4. Direct a discussion with each group sharing some of their observations. Help the students see that one way the block and golf ball are the same is that they are both examples of matter. Share the overhead transparency on the properties of matter. Place emphasis on the idea that the block of wood and the golf ball take up space and have mass.

5. Distribute the student book. Tell the students to read the information on mass in the student mini-book. Make sure they have a beginning understanding of mass before continuing with the lesson. Have the students give examples

around the room of objects that have mass and take up space.

6. Hold up a cup of water and ask, "Does this have mass and take up space?" Distribute a cup to each group and ask the students to find the mass of the cup and water. Ask them, "How can you find the volume of the water in the cup?" Ask them to find the volume of the water in the cup using the graduated cylinder. Direct the students to compare the cup of water to the block of wood and the golf ball. Ask the students, "How are they the same? How are they different?"

7. Lead the class in a discussion by allowing each group to share some of their observations.

8. Ask the students, "Do solids and liquids have volume and take up space?" [Yes] "How could you prove that they do?"[through measurement] Show the students a deflated balloon. Ask them, "Does this balloon have mass and does it take up space?"

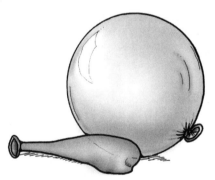

9. Distribute two balloons to each group. Tell the students to place one balloon in the each of the pans of the balance. Direct the students to zero the balance once the balloons are in the pans. Have the students fully inflate one of the balloons. Ask, "What has been added to the balloon to change its mass?"

10. Direct the students to place the inflated balloon back into the pan. Have them carefully observe the pointer on the balance. They should observe the rocking motion of the pointer as it is coming

back in balance. The pointer will rock more to the side that has the inflated balloon. Ask, "What does the balance show us about the mass of the inflated balloon?" [The air has some mass. When you add the mass of the air to the mass of the balloon, the total mass increases. The increase is very small, but the balance helps us see the difference even though we cannot measure the difference with the instruments we are using.]

11. Ask the students to find examples of other types of matter in the room. Find the mass of each and find the volume of each.

Connecting Learning

1. In your own words, define matter.
2. What three states of matter did we explore today? [solid, liquid, and gas]
3. What state of matter was the easiest to find? Why do you think it was easiest to find this state of matter?

4. Why is it difficult to find the mass and volume of some state of matter?
5. How are mass and volume different?
6. Give an example of something that is not matter. [Any type of energy would be an example for instance light is not matter.] How do you know that it is not? [Light does not have mass and it does not take up space]

Evidence of Learning

1. Listen for correct explanation that matter has mass and it takes up space.
2. Look for both mass and volume recordings on data sheet for matter search.

* Reprinted with permission from *Principles and Standards for School Mathematics*, 2000 by the National Council of Teachers of Mathematics. All rights reserved.

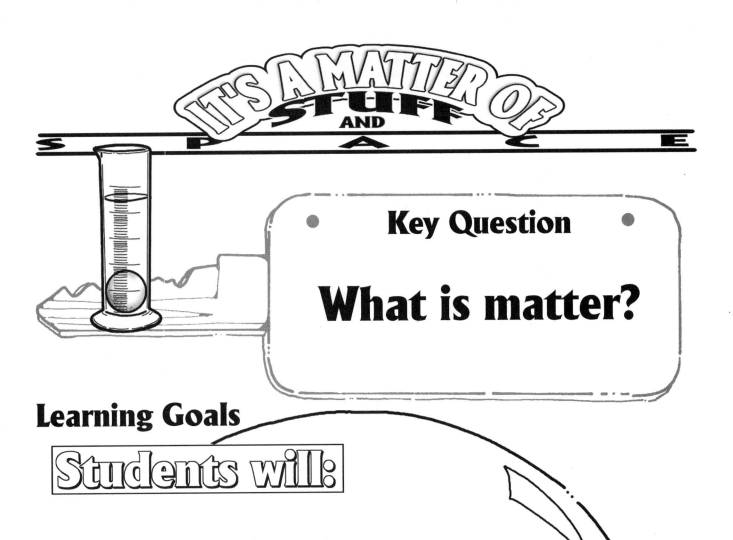

IT'S A MATTER OF STUFF AND SPACE

Key Question

What is matter?

Learning Goals

Students will:

1. **define matter as anything that has mass and takes up space, and**

2. **identify examples of matter found in their classroom.**

Matter is defined as anything that
- has mass and
- takes up space.

Mass describes the amount of material in the matter. Mass is measured with a balance in grams.

Space describes the amount of volume of the matter. The volume or space of matter can be calculated with formulas or through a process known as displacement.

	volume	mass
wood block		
golf ball		
water		

List all the characteristics.

How are they alike?

wood block   golf ball   water

How are they different?

wood block   golf ball   water

20
19
18
17
16
15
14
13
12
11
10
9
8
7
6
5
4
3
2
1

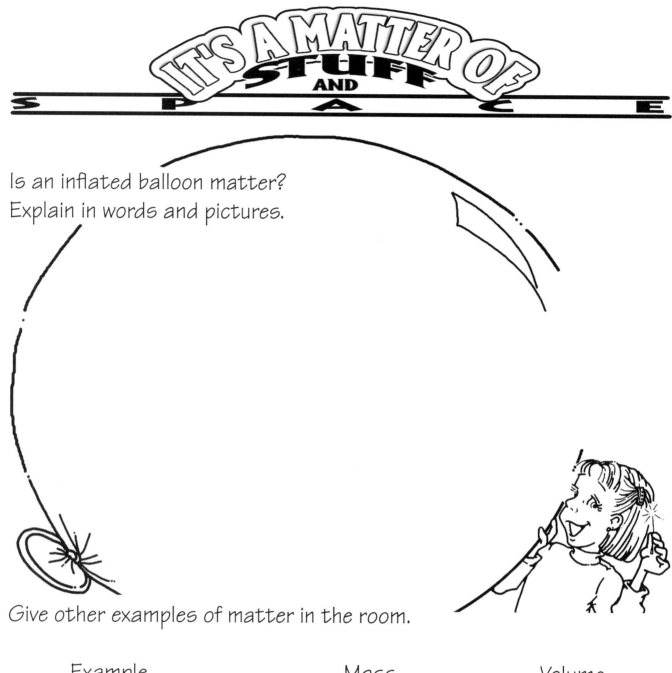

IT'S A MATTER OF STUFF AND SPACE

Is an inflated balloon matter?
Explain in words and pictures.

Give other examples of matter in the room.

Example	Mass	Volume

18

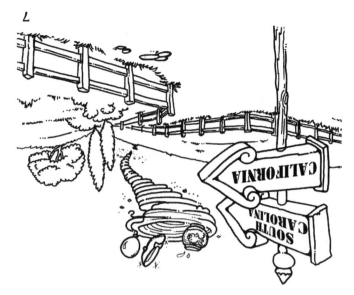

Most materials can go from one state to another. That doesn't meaning traveling from South Carolina to California. It means it can change from one phase to another, such as a solid to a liquid. A change in temperature or pressure can cause a change in state.

The differences between a solid, liquid, and gas are very apparent when you look at and touch them. Most solids have specific shapes. They do not need a container to keep their shapes. If you put a liquid in a container, it will normally stay in the container and take the shape of the container. A gas will also take the shape of its container, but it will not normally stay in the container.

A good example of how matter can change from one state to another is what happens to water. Water is normally a liquid at room temperature, but if you cool it below 0° C, it will freeze and become the solid we call ice. If you heat water above 100° C, it will boil and turn into a gas we call steam.

You may know that matter on Earth exists in one of three phases or states: solids, liquids, or gases. The question we will address in this book is why.

8

1

The molecules in liquids move slower than molecules in a gas. The attraction between the molecules is greater than their movement that pulls them apart. This mild binding force holds the material together in the form of a liquid.

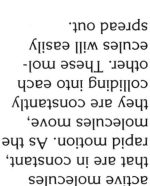

Gases have very active molecules that are in constant, rapid motion. As the molecules move, they are constantly colliding into each other. These molecules will easily spread out.

When a liquid is cooled even more, the movement of the molecule slows down so much that the attractive forces between the molecules form the structure of a solid.

But what makes a substance a solid, a liquid, or a gas? To begin to understand this, you have to know some things about molecules. All substances are made up of molecules. Molecules are too small to be seen without the aid of very powerful microscopes. The other things you must understand are about spacing and the movement between the molecules. Molecules are in constant motion and they have forces of attraction between them.

6

3

Connecting Learning

1. In your own words, define matter.

2. What three states (phases) of matter did we explore today?

3. What state (phases) of matter was the easiest to find? Why do you think it was easiest to find this state of matter?

4. Why is it difficult to find the mass and volume of some states (phases) of matter?

5. How are mass and volume different?

6. Give an example of something that is not matter. How do you know that it is not?

7. What are you wondering now?

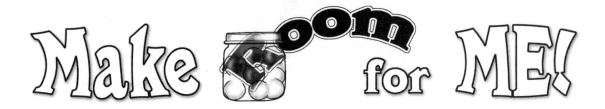
Make Room for ME!

Topic
Matter

Key Question
How does matter occupy space?

Learning Goals
Students will:
1. combine different liquids and solids to discover that these materials have spaces between their particles, and
2. measure and combine the volumes of different substances.

Guiding Documents
Project 2061 Benchmark
- *All matter is made up of atoms, which are far to small to see directly through a microscope.*

NCTM Standards 2000*
- *Understand such attributes as length, area, weight, volume, and size of angle and select the appropriate type of unit for measuring each attribute*
- *Solve problems that arise in mathematics and in other contexts*

Math
Measurement
   volume
Graphs

Science
Physical science
   chemistry
      matter

Integrated Processes
Observing
Comparing and contrasting
Predicting
Inferring
Generalizing

Materials
For each group:
   popcorn (unpopped)
   sand
   large marbles
   2 graduated cylinders, 100 mL
   baby food jar
   rock salt
   water
   isopropyl alcohol, commonly sold as rubbing alcohol (see *Management 1*)
   eyedropper
   paper or plastic cups for liquid waste
   safety goggles

Background Information
One of chemistry's greatest achievements was to show that all the matter found in nature is built from about one hundred *elements*. For example hydrogen, oxygen, carbon, gold, and helium are all elements. The smallest amount of an element that can exist is an *atom*. Chemists can mix, heat, soak in acid, or use other chemical techniques for changing matter, but they cannot break matter down into pieces smaller than an atom of each element. Physicists use *physical* (not *chemical*) techniques for breaking atoms into electrons, protons, and other fundamental particles.

Atoms can combine into *molecules*. For example, one oxygen atom combines with two hydrogen atoms to form *one molecule* of water. Three carbon atoms combine with eight hydrogen atoms and one oxygen atom to form *one molecule* of isopropyl alcohol.

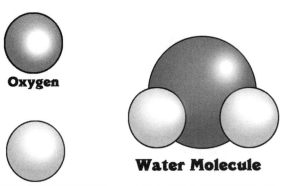

Oxygen

Hydrogen

Water Molecule

No two atoms or molecules can occupy the same space at the same time. The molecules of water and alcohol are different in size and shape and therefore, when combined, the smaller water molecules fit between the larger isopropyl alcohol molecules. Students will be amazed when they find that the volumes of the two liquids are not additive.

Management

1. If possible, order reagent grade isopropyl alcohol. It contains less water than rubbing alcohol. WARNING: **Do not use methyl alcohol!**
2. If necessary, review how to read a graduated cylinder. Remind students that they will need to read the level of the liquid at the bottom of the meniscus.

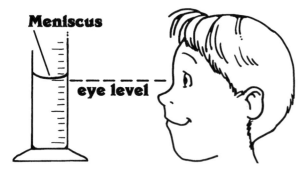

3. Help students make accurate volume measurements. Sometimes students will find it helpful to use an eyedropper to add the last few drops.
4. In *Part Three*, other materials may be substituted, such as ball bearings, rock salt, sugar, walnuts, etc., for those materials listed on the student sheets.
5. Sand may be eliminated from *Part Three* if separating it from the combined mixture presents a management problem.
6. Instruct students to pour the popcorn through cupped hands or a funnel to prevent the accidental scattering of popcorn kernels.
7. Distribute cups for students to pour their waste liquids into for later disposal.

Procedure

Part One

Caution: Warn students to not come in contact with the alcohol and be sure that they wear goggles!

1. Instruct the students to pour 50 mL of water into one graduated cylinder. Monitor the students' measurements and assist as necessary.
2. Have them pour another 50 mL of water into a second graduated cylinder.
3. Ask them to predict the total volume when the two liquids are combined and record their predictions.
4. Instruct them to combine the two liquids and record the combined volume.
5. Tell them to shade the graduated containers on the student sheet.
6. Direct the students to empty their graduated containers into the waste-liquid cups.
7. Allow the students time to share and discuss their results with their neighbors.
8. Ask the students to repeat these steps, only this time using 50 mL of isopropyl alcohol in each

graduated container. Instruct the students to save this alcohol in a cup for later use (step 10).
9. Give the students time to share and discuss their results with their neighbors.
10. Instruct the students to repeat the process a third time using 50 mL of water in one graduated container and 50 mL of isopropyl alcohol in the other graduated cylinder.
11. Have the students share and discuss their results with their neighbor.
12. Give the students time to write their explanations for the results of the water-alcohol combination on their sheets.

Part Two

1. Instruct the students to attach a piece of masking tape vertically to a baby food jar. Then tell them to half-fill a dry baby food jar with rock salt.
2. Next, have them add water to a level two or three cm above the level of the salt and to then mark the level of the liquid on the strip of tape.
3. Tell the students that they will soon be asked to shake their containers, but before they do, they need to predict and record where the level of the water and salt mixture will be after shaking: *on the line, above the line, below the line.*
4. Warn them to cover the container with the lid of the jar before shaking. Inform them that they will need to shake their containers for about one minute.
5. After at least one minute of shaking, tell the students to measure and record the new level of the water and salt mixture. (Results should show a decrease in the level.)

Part Three

1. Using a 1/3 or 1/4 cup measuring cup or a jar lid as a *scoop*, have the students predict and record how many scoops it will take to fill the container used in *Part Two* with marbles.
2. Instruct the students to use their scoops to "fill" their containers with marbles and to record on their activity sheets the number of scoops added.
3. Ask the students to predict the number of scoops of popcorn that can be added to their container to make it "full." On their activity sheets, have them record the actual number of scoops of popcorn they added to their container.
4. Tell the students to predict the number of scoops of sand that can be added to their container to make it "full." Have them record the actual number of scoops of sand they added to their container.
5. Have students make a drawing of their container and its contents.
6. Instruct students to graph the actual measurements for each item they added to their containers.
7. Use a sieve to reclaim the materials.

23

Connecting Learning

Part One

1. What are some of the characteristics of *matter?*
2. How do you explain the results of adding 50 mL of water to 50 mL of water?
3. How do you explain the results of adding 50 mL of alcohol to 50 mL of alcohol?
4. How do you explain the results of adding 50 mL of water to 50 mL of alcohol?

Part Two

1. What effect does shaking have on the mixture of salt and water?
2. How do you explain the results of combining water with rock salt and then shaking the mixture?

Part Three

1. Discuss why you do or do not think the container is full at the conclusion of *Part Three.*
2. What do you think happens, over time, to a large bag of potato chips?
3. Compare the volume of a bag of potato chips with a canned variety of potato chips. Count and compare the number of chips in a large bag of potato chips with the number of chips in a can. Discuss your findings.
4. Summarize what you have learned from this activity.
5. What are you wondering now?

Extensions

Part Three

1. After each group completes the investigation, compare the percent of each item that was added to the container.
2. Try filling the container with the same materials but in a different order. What happens?

 * Reprinted with permission from *Principles and Standards for School Mathematics,* 2000 by the National Council of Teachers of Mathematics. All rights reserved.

Make Room for ME!

How does matter occupy space?

Learning Goals

STUDENTS WILL:

1. combine different liquids and solids to discover that these materials have spaces between their particles, and

2. measure and combine the volumes of different substances.

Make Room for ME!

Part 1

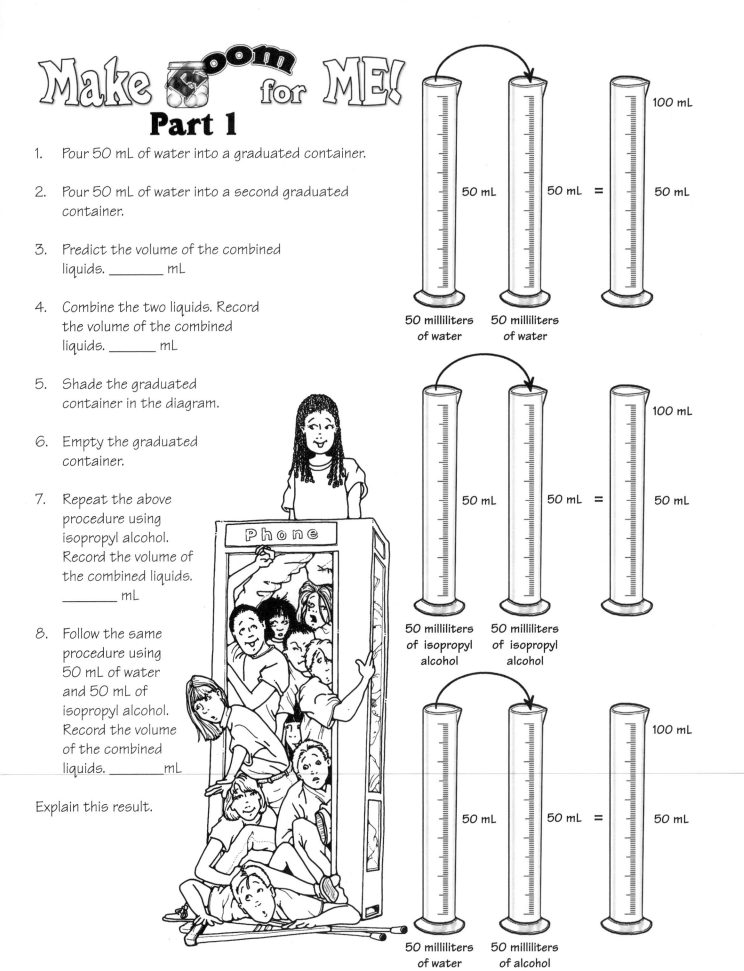

1. Pour 50 mL of water into a graduated container.

2. Pour 50 mL of water into a second graduated container.

3. Predict the volume of the combined liquids. _____ mL

4. Combine the two liquids. Record the volume of the combined liquids. _____ mL

5. Shade the graduated container in the diagram.

6. Empty the graduated container.

7. Repeat the above procedure using isopropyl alcohol. Record the volume of the combined liquids. _____ mL

8. Follow the same procedure using 50 mL of water and 50 mL of isopropyl alcohol. Record the volume of the combined liquids. _____mL

Explain this result.

100 mL

50 mL   50 mL   =   50 mL

50 milliliters   50 milliliters
of water   of water

100 mL

50 mL   50 mL   =   50 mL

50 milliliters   50 milliliters
of isopropyl   of isopropyl
alcohol   alcohol

100 mL

50 mL   50 mL   =   50 mL

50 milliliters   50 milliliters
of water   of alcohol

Make Room for ME! Part 2

1. Fill a graduated container 1/3 full of rock salt.

2. Add water to a level 2-3 cm above the level of the rock salt. Mark the level of water and salt mixture on the masking tape.

3. Predict the level after shaking.
   _____ above the line
   _____ on the line
   _____ below the line

baby food jar

jar lid

masking tape

water

salt

4. Cover the container with the jar lid. Shake the container for at least one minute.

5. Record the results.
   _____ above the line
   _____ on the line
   _____ below the line

How do you explain the results?

27

Make oom for ME! Part 3

1. Using the jar lid as a "scoop," predict how many scoops it will take to fill a baby food jar with marbles.

   Fill the container with marbles. Record the number of scoops. Leave the marbles in the jar.

2. Predict how many scoops it will take to fill the container with popcorn.

   Add the popcorn to the container. Record the number of scoops. Leave the popcorn in the container.

3. Predict how many scoops it will take to fill the container with sand.

   Add the sand to the container. Record the number of scoops.

Prediction

_____ Scoops

Measurement

_____ Scoops

Prediction

_____ Scoops

Measurement

_____ Scoops

Prediction

_____ Scoops

Measurement

_____ Scoops

4. Make a drawing of your container and its contents.

5. Graph the actual measurements for each item added to the container.

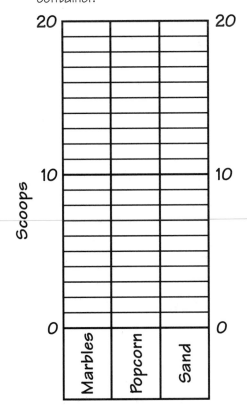

Make room for ME!

Connecting Learning

Part One

1. What are some of the characteristics of *matter*?

2. How do you explain the results of adding 50 mL of water to 50 mL of water?

3. How do you explain the results of adding 50 mL of alcohol to 50 mL of alcohol?

4. How do you explain the results of adding 50 mL of water to 50 mL of alcohol?

Part Two

1. What effect does shaking have on the mixture of salt and water?

2. How do you explain the results of combining water with rock salt and then shaking the mixture?

Connecting Learning

Part Three

1. Discuss why you do or do not think the container is full at the conclusion of *Part Three*.

2. What do you think happens, over time, to a large bag of potato chips?

3. Compare the volume of a bag of potato chips with a canned variety of potato chips. Count and compare the number of chips in a large bag of potato chips with the number of chips in a can. Discuss your findings.

4. Summarize what you have learned from this activity.

5. What are you wondering now?

Metal Matters!

Topic
Properties of Matter

Key Question
What types of metals are attracted to a magnet?

Learning Goals
Students will:
1. classify metals based on their attraction to a magnet; and
2. identify iron, nickel, cobalt and manganese as those having magnetic properties.

Guiding Documents
Project 2061 Benchmark
- *Without touching them, a magnet pulls on all things made of iron and either pushes or pulls on other magnets.*

NRC Standard
- *Objects are made of one or more materials, such as paper, wood, and metal. Objects can be described by the properties of the materials from which they are made, and those properties can be used to separate or sort a group of objects or materials.*

Science
Physical science
  matter
  magnetism

Integrated Processes
Observing
Comparing and contrasting
Classifying
Predicting
Inferring
Generalizing
Applying

Materials
For each group:
  wooden ruler (see *Management 3*)
  tape (see *Management 3*)
  ring magnet (see *Management 3*)
  brass house key
  metal paper clip
  metal spoon
  aluminum can
  steel safety pin

  piece of copper
  nickel
  quarter
  tin can
  aluminum nails
  brass screws
  iron nails

For each student:
  metal mini-book

Background Information
All matter displays a certain measurable amount of magnetic force. Highly sophisticated scientific equipment is needed to measure the force at very low levels. Ferromagnetic metals are highly magnetic. The most common ferromagnetic metals are iron, cobalt, and nickel or combinations of these with other materials. You don't find too many products made of pure iron, but you do find many products made of steel. Since steel has a lot of iron in it, steel is attracted to a magnet. Other common ferromagnetic metals are cobalt and nickel, or combinations of these two materials.

To make any ferromagnetic metal into a magnet, the metal to be magnetized is placed into a device called a magnetizer. The magnetizer has a powerful electric current traveling through coils of wire. The electricity running through the wires creates a strong magnetic field that magnetizes the metal by aligning clusters of atoms that have magnetic properties. These clusters are called magnetic domains. Once these clusters have been aligned the metal has magnetic properties.

Management
1. The metals suggested can be easily obtained from your home as well as a local hardware store. With a little advance planning, you can solicit some of these items by sending a letter of request to parents.
2. Be sure there are no rough edges on the metals that could cut the students.
3. The students will need to construct a magnetic wand by attaching a ring magnet to the end of the wooden or plastic ruler with the tape.

Procedure

1. Ask the *Key Question* and state the *Learning Goals*.
2. Distribute the materials to the students.
3. Direct them to use the sorting sheet for their prediction on which metals will be attracted to a magnet and which will not be attracted. Have them record the predictions on the data sheet.
4. Demonstrate how to attach the magnet to the ruler with tape.
5. Tell the students to test their predictions and record what they find.
6. Discuss with the students their findings, and direct them to read the student mini-book on magnetic properties.
7. Ask the students to label the items on the student sheet that are composed of a ferrous metal.

Connecting Learning

1. What surprised you while doing this activity?
2. What types of metals did you find to be highly attracted to magnets?
3. Why do you think most keys are brass, a non-magnetic metal? [A magnetic metal could cause problems with the locking mechanisms in a lock.]
4. How have magnetic properties of metals been applied for use in our everyday lives?
5. What other properties of matter do you think you might be able to explore?
6. What are you wondering now?

32

METAL MATTERS!

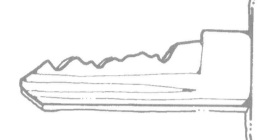

•Key Question•

What types of metals are attracted to a magnet?

Learning Goals

Students will:

1. classify metals based on their attraction to a magnet; and

2. identify iron, nickel, cobalt, and manganese as those having magnetic properties.

Metal Matters!

List the objects here. Predict whether they will be attracted to a magnet or not.

_____   _____   _____

_____   _____   _____

_____   _____   _____

_____   _____   _____

_____

Use this space to sort your objects.

☆ ☆ ☆ *Magnetic* ☆   ☆   ☆   ☆ *Non-Magnetic* ☆

After testing your objects, record your results. Are they attracted to a magnet or not attracted to a magnet?

_____   _____   _____

_____   _____   _____

_____   _____   _____

_____   _____   _____

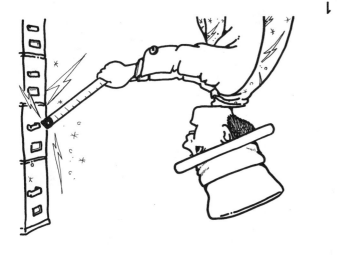

You found out that certain metals, but not all, are magnetic. Metals that display magnetic properties are called ferromagnetic.

Whose held the least? How could you find the mode for the class? What about the median? What about the mean?

1

12

Iron is the most common ferromagnetic element. The atomic symbol for iron is Fe.

The stronger your magnet, the more paper clips it should hold! How many could yours hold? Whose held the most?

2

11

Something else to think about: Did you know that if your magnet is strong enough, the magnetic field of the magnet could be extended all the way through the object so that another magnetic object can stick to the first?

The other known elements that are capable of producing a magnetic field are cobalt and nickel.

Try this with your magnetic wand. Attract a large paper clip to your magnet. Once it is stuck, see if a small paper clip will stick to the bottom of the large paper clip.

Where do the symbols for the name of the elements appear to come from?

The Latin word for iron is ferrum.

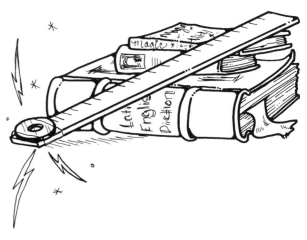

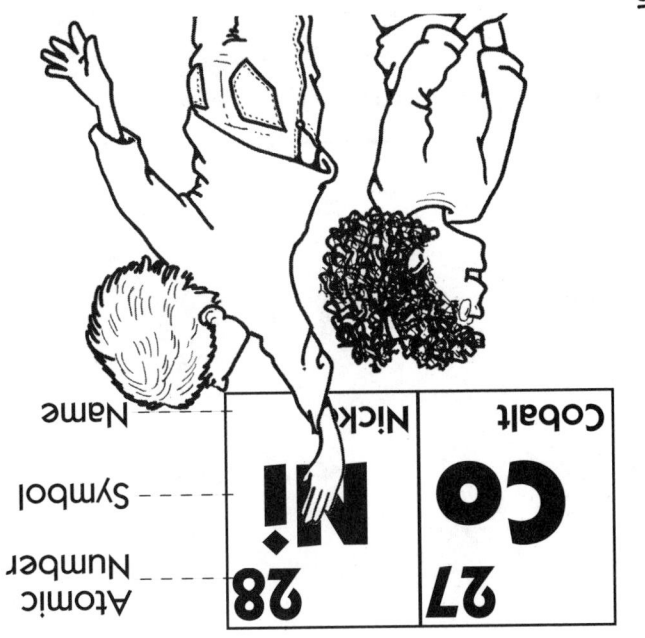

Name

Symbol

Atomic Number

Cobalt Co 27

Nickel Ni 28

Cobalt's atomic symbol is Co and nickel's atomic symbol is Ni.

The nickel coin is composed of 25 percent nickel and 75 percent zinc.

The nickel we use in our money system is an alloy. An alloy is two or more metals combined. It can also be a metal and a non-metal combination.

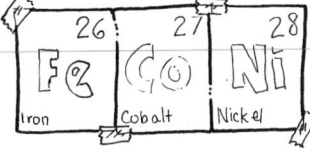 

26 Fe Iron   27 Co Cobalt   28 Ni Nickel   30 Zn Zinc

Metal Matters!

Connecting Learning

1. What surprised you while doing this activity?

2. What types of metals did you find to be highly attracted to magnets?

3. Why do you think most keys are brass, a non-magnetic metal?

4. How have magnetic properties of metals been applied for use in our everyday lives?

5. What other properties of matter do you think you might be able to explore?

6. What are you wondering now?

Chromatographic Circles

Topic
Separating a Mixture

Key Questions
1. How do the inks in black pens compare?
2. What colors are in black ink?

Learning Goals
Students will:
1. identify the colors that make up black ink in a water-soluble pen, and
2. identify how chromatography can be used to separate a mixture.

Guiding Documents
Project 2061 Benchmarks
- *Materials may be composed of parts that are too small to be seen without magnification.*
- *When a new material is made by combining two or more materials, it has properties that are different from the original materials. For that reason, a lot of different materials can be made from a small number of basic kinds of materials.*

NRC Standard
- *A substance has characteristic properties, such as density, a boiling point, and solubility, all of which are independent of the amount of the sample. A mixture of substances often can be separated into the original substances using one or more of the characteristic properties.*

Science
Physical science
    matter

Integrated Processes
Observing
Comparing and contrasting
Communicating
Predicting
Collecting and recording data
Classifying
Applying

Materials
For the class:
    clock or timing device

For each group of four students:
    4 standard paper clips
    4 different black markers (see *Management 4*)
    4 round coffee filters (see *Management 2*)
    4 Styrofoam plates
    ruler

Background Information
Chromatography is a method used to separate the components of a mixture. The name comes from the Greek words *chroma*, which means color and *graph*, which means to write. This activity uses a process called paper chromatography to separate the black ink in water-soluble pens. A spot of ink is placed on a strip of filter paper. The strip is then suspended so that just the tip of the paper is in the water. The spot of ink is not in the water; it is just above the water. The water moves up the strip by the process of capillary action (like a wick) and is able to dissolve the components in the ink spot.

The different components in the ink have different solubilities in water. That means they have different levels of attraction for the moving water and the stationary pieces of filter paper. The components that have a greater attraction for the water (are more soluble), separate more readily from the mixture, and travel up the paper first. The components that have less attraction for water are less soluble and travel up the filter paper slower. Given enough time and enough difference in the solubilities of the components, the components can be completely separated. The chemicals that make up colors are called pigments. Black ink is a solution, a mixture of several pigments. Different companies use different pigments to make the black ink we see in markers. Water-soluble components also differ from permanent ink components.

Management
1. This activity works best for student groups of four.
2. Iron the coffee filters before distributing them. Use a medium setting and iron several filters at a time.
3. The cotton balls should be wet but not dripping wet.
4. Make sure to have four different brands of water-soluble pens. Each student within a group will select one of the four pens.

Procedure

1. Ask the *Key Questions* and discuss the *Learning Goals.*
2. Distribute the coffee filters and paper clips.
3. Demonstrate how to fold the coffee filter in half then in half again creating a quarter of a circle. Unfold the coffee filter and locate the midpoint of the coffee filter from the two folds.

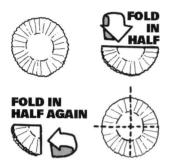

4. Demonstrate how to use the paper clip and two pencils as a compass to draw a circle from the midpoint of the coffee filter. Place the paper clip so that you can see the midpoint of the coffee filter inside one of the loops. Place the point of one pencil on the midpoint of the coffee filter. Place the second pencil through the loop at the other end of the paper clip. Hold the pencil that is on the midpoint in place and draw a circle using the paper clip to hold the second pencil the same distance from the midpoint as you are drawing the circle.

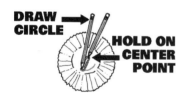

5. Tell the students to use their selected pens to draw a design on the coffee filter. Make sure the students draw their designs outside the circles they just drew with the paper clips. Students may want to make labels for the brands of pens they used. This will help as they describe their observations.

6. Distribute the Styrofoam plates and the wet cotton balls.
7. Tell the students to place the wet cotton ball in the center of their plates. The cotton balls should not be dripping water.
8. Have them place the coffee filter on top of the cotton ball so that the circle they drew is centered over the cotton ball.
9. Invite students to record their observations. Tell them to record what they see happening—which colors and when they appeared. Encourage them to record which colors appeared to move the farthest and which appeared to have moved the least.
10. Have students compare their chromatograms with the others in their groups.
11. Read and discuss the student information sheet on chromatography.

Connecting Learning

1. What happened to the designs on the filter paper?
2. What colors did you see in your ink?
3. Did the same colors appear in all the chromatograms? Explain.
4. What evidence did you observe in your chromatogram that would lead you to believe that one color in your ink was more soluble than another? [The further the ink travels away from the water source, the more soluble it is.]
5. Was the same color always the least or most soluble? Explain.
6. What could you tell someone to convince them that water-soluble black ink is a mixture? [Black ink would be a mixture because different pigments are present.]

Evidence of Learning

Listen for discussion based on observations from the activity as well as information gained from the reading passage. Students should be able to tell you that chromatography is the process by which a mixture can be separated.

Chromatographic Circles

Key Questions

1. How do the inks in black pens compare?
2. What colors are in black ink?

Learning Goals

Students will:

1. identify the colors that make up black ink in a water-soluble pen, and

2. identify how chromatography can be used to separate a mixture.

What did you observe?

Record the order of the colors you see from first to last.

Color in your final design.

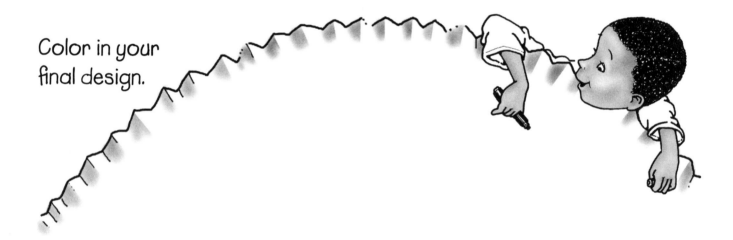

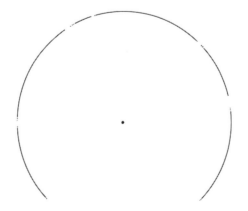

44

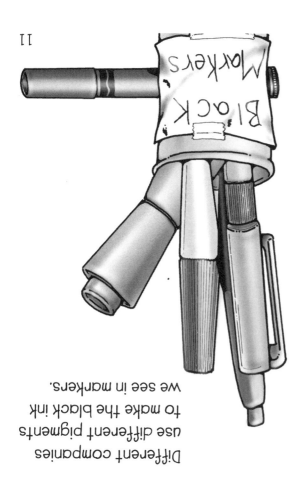

Different companies use different pigments to make the black ink we see in markers.

The name comes from the Greek words chroma which means color, and graph which means to write.

Water-soluble pigments are different from permanent ink pigments.

Chromatography is one method used to separate the components of a mixture.

12

1

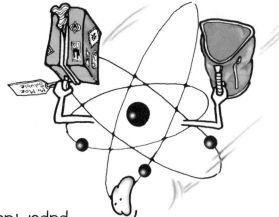

The pigments that have a greater attraction for the water are more soluble and spend more time in the water. These are the colors that travel up the paper faster.

The different pigments in the ink have different solubility in water. That means they have different levels of attraction for the moving water and attraction to the filter paper.

The spot of ink is not in the water; it is just above the water. The water moves up the strip by the process of capillary action (like a wick) and is able to dissolve the components in the ink spot.

The pigments, which have less attraction for water, are less soluble, and travel up the filter paper more slowly.

8

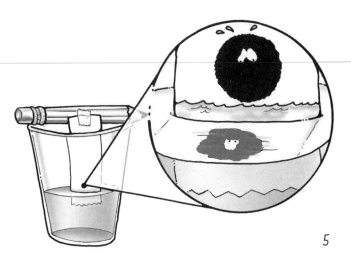

5

3

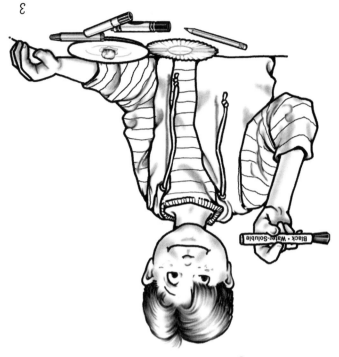

We used paper chromatography to separate the black inks in our investigation.

10

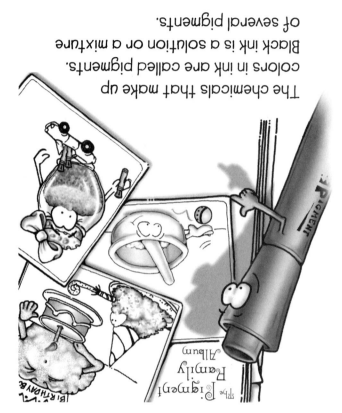

The chemicals that make up colors in ink are called pigments. Black ink is a solution or a mixture of several pigments.

Paper chromatography uses strips of filter paper. The coffee filter paper you used is similar to the type of filter paper used in traditional paper chromatography. A spot of ink is placed on the strip. The strips are then suspended so that just the tip of the paper is in the water.

4

If you have enough time and there are differences in the solubility of the pigments, the pigments can be completely separated.

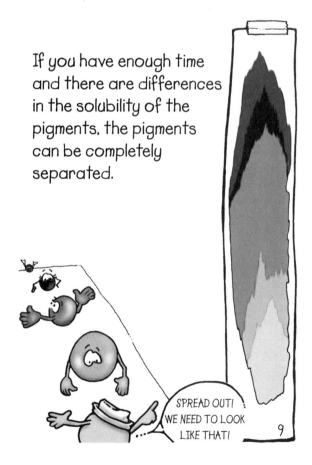

SPREAD OUT! WE NEED TO LOOK LIKE THAT!

9

Chromatographic Circles

Connecting Learning

1. What happened to the designs on the filter paper?

2. What colors did you see in your ink?

3. Did the same colors appear in all the chromatograms? Explain.

4. What evidence did you observe in your chromatogram that would lead you to believe that one color in your ink was more soluble than another?

5. Was the same color always the least or most soluble? Explain.

6. What could you tell someone to convince them that water-soluble black ink is a mixture?

7. What are you wondering now?

Messing with Mixtures

Topic
Separating Mixtures

Key Question
How can the ingredients of a mixture be separated?

Learning Goals
Students will:
1. identify what a mixture is; and
2. explore how filtration, settling, and evaporation can be used to separate a mixture.

Guiding Documents
Project 2061 Benchmarks
- *No matter how parts of an object are assembled, the weight of the whole object made is always the same as the sum of the parts; and when a thing is broken into parts, the parts have the same total weight as the original thing.*
- *Heating and cooling cause changes in the properties of materials. Many kinds of changes occur faster under hotter conditions.*
- *Materials may be composed of parts that are too small to be seen without magnification.*
- *When a new material is made by combining two or more materials, it has properties that are different from the original materials. For that reason, a lot of different materials can be made from a small number of basic kinds of materials.*
- *When liquid water disappears, it turns into a gas (vapor) in the air and can reappear as a liquid when cooled, or as a solid if cooled below the freezing point of water.*

NRC Standard
- *A substance has characteristic properties, such as density, a boiling point, and solubility, all of which are independent of the amount of the sample. A mixture of substances often can be separated into the original substances using one or more of the characteristic properties.*

NCTM Standards 2000*
- *Solve problems that arise in mathematics and in other contexts*
- *Understand such attributes as length, area, weight, volume, and size of angle and select the appropriate type of unit for measuring each attribute*

Math
Measurement
  mass

Science
Physical science
  chemistry
    mixtures

Integrated Processes
Observing
Predicting
Collecting and recording data
Comparing and contrasting
Applying

Materials
For each group of students:
  2 plastic cups, 8 oz.
  small strainer (fine mesh)
  balance
  masses

For the class:
  1 cup plaster of paris
  1 cup sugar
  1 cup wood shavings or sawdust
  4 cups water
  medium-sized bucket
  large stirring spoon
  newspapers
  saucepan with lid, optional
  hot plate, optional

Background Information
Most of the substances we know are made of different combinations of elements. If they are formed with a definite, set proportion of substances that are strongly linked chemically, they are called *compounds*. Water is an example of a compound; so is table salt. If the ingredients are mixed (often with variable proportions) and not bonded chemically, the combination is called a *mixture*. Powdered drink mix, salad dressing mix, and playground dirt are familiar examples of mixtures.

Pure substances can consist of combinations of both elements and compounds. In this activity students will use plaster of paris, a substance that comes from a rock called gypsum. Gypsum is composed of calcium sulfate, a compound of calcium, sulfur and oxygen, and water. Two of the other substances used in this activity, water and sugar, are also compounds. The fourth ingredient, sawdust (or wood shavings), is a mixture.

Mixtures usually can be separated into their original ingredients. For example, unclean water is a mixture of water and other things such as mud and pollutants. Certain tools and technology enable us to make unclean water usable again. Among the common methods used are:

- *filtration* (straining the material through some kind of screen to remove the larger particles),
- *settling* (allowing the more dense materials to sink to the bottom so the less dense material can be separated from it),
- *evaporation* (changing the liquid to a vapor which then escapes into the atmosphere), and
- *distillation* (the heating of a liquid to a vapor, but this time the vapor is caught and condensed to liquid form again).
- In this activity, the students will experience some of these methods used to separate mixtures.

Management

1. This activity will take two class periods (one long, one short) plus observation over a week or more.
2. The proportions listed will make 10-12 portions of mixture. If you need more, double the recipe.
3. Caution the students not to taste the mixture or pour it down the sink. Appropriate disposal of these materials is to contain them and place them in the dumpster.
4. After mixing the ingredients in the bucket, keep stirring right up to the moment you pour them into individual cups; otherwise much of the plaster will sink to the bottom of the bucket and not be included in the student portions.
5. Do not show or give the students tools such as the strainers until such time as they determine they are needed.
6. Designate a place where the cups of mixture can be left without being in danger of getting knocked over. A warm place such as a sunny windowsill will help speed up the process of evaporation.

7. If you have an insect problem in your classroom, you will need to protect the cups of sugar water. Placing the cups in a pan with a few inches of water will help discourage ants and other crawling insects. Net or screen over the top of the cups will help keep out flies.

Procedure

(The following is offered for those students who are ready for more independent work.)

Open-ended:
Give students with experience in working with mixtures a sample of this mixture and ask them to come up with their own solution to the problem of separating it into its original ingredients. Direct them to carry out their plan, recording and analyzing their processes and results as they go.

Guided Instruction
Day One

1. Tell the class that you are going to create a very interesting and messy mixture. With the class watching and participating, spread out newspapers and put out a large container or bucket. Dump in the plaster of paris and slowly add water so as not to create a cloud of plaster dust that may get in your eyes. Stir well until both ingredients are mixed. Add the sawdust and stir again. Finally, add the sugar and keep stirring. Caution the students not to try to taste the mixture because the plaster and sawdust are indigestible.
2. Distribute a cup of the mixture (and a folded section of newspaper to put under it) to each group of students. Keep stirring between servings to be sure that none of the mixture settles.
3. Have each group of students find the mass of both the cup of mixture and the empty cup and record.
4. Ask the students to think of ways to separate the ingredients of their mixture to get back the four original substances. Have them brainstorm the ideas and make a list. If necessary, guide the discussion to the observation that one of the ingredients (sawdust) is floating and thus may be easier to remove. Ask the students for suggestions as to how to remove the sawdust and list different ideas.
5. Explain to the students that one way used in industry to separate mixtures is called *filtration*. Some may connect this idea to filters they have seen used in coffeepots or vacuum cleaners. Give each group a strainer and a second cup and have them filter out the sawdust by pouring the contents through the strainer into the empty cup. They may need to pour the contents of the cup back and

forth several times to remove as much sawdust as possible. Have the students drain the soggy sawdust into the cup of mixture. Tell them to put the sawdust into the other cup, find its mass, and record. Have them set the sawdust aside on the newspaper and save.

6. Ask the students to identify the ingredients that are left in the cup and discuss how they could be further separated. Record all the suggestions. Encourage them to keep thinking about this problem until the next day. Tell each group to leave the cup and sawdust pile on the newspaper in a designated place where they will not be disturbed. Direct them to tidy their areas and to rinse the extra cup for later use. Be very careful not to let anyone pour any of the mixture in the sink.

Day Two
1. Before the students bring their cups and newspaper to their desks, ask if they have had any more ideas about how to separate the remaining ingredients. List any new ideas.

2. Have the students bring the mixture cups back to their desks **very carefully**. Tell them not to stir it and to try not to disturb the contents any more than necessary. The students should notice that the plaster of paris has settled to the bottom. This will make it possible for them to pour off the sugar/water into the second cup and leave the plaster in the bottom of the first cup. This process is called *settling* and is also used in industry to get rid of some of the denser, undesirable materials.

3. When the students have poured off the sugar water, have them find the mass of the cup with the plaster in it, record, and set it aside. Direct them to observe the sugar water. Again, ask for ideas as to how these remaining ingredients could be separated from each other. Make a list of the suggestions and discuss how they might be carried out.

4. The last stage of separation involves either *evaporation* or *distillation*. Both of these processes involve turning the water to vapor and leaving the sugar behind.

    Evaporation: Have the students put their cups in a safe place, preferably warm, and check on them every day until the evaporation process is complete (the water vapor has escaped to the atmosphere) and only a sugar residue is left. This can take anywhere from a few days to a few weeks depending on how much liquid is in the cup and how warm the surrounding temperature is.

    Distillation: Heat the sugar water mixture in a saucepan over a hot plate. Trap some of the escaping water vapor in the lid of the saucepan, let it cool and condense, and pour it off the lid into a cup to show the water to the children. If

you do not use a lid, you are merely speeding up the process of evaporation. In either case, you will end up with sugar residue in the pan. Be careful not to burn it!

5. When the water has evaporated, have the students examine the residue in the cup. Have them find the mass of the cup of sugar, record, and put it with the plaster and sawdust.

6. Have the students look at the pile of sawdust, the cup of plaster, and the cup of sugar residue and review the various separation processes they used. Ask them if the mass of the mixture's components totals the mass they found at the beginning of the activity. Discuss the differences. Challenge them to determine how much water in the original mixture has now evaporated.

Connecting Learning
Day One
1. What do you observe about this mixture? How do you know that it is a mixture and not a compound?
2. What different ways can you think of to separate this mixture back into its original ingredients? Which idea do you think will work the best? Why do you think so?
3. What tools can you think of that might help you and your team separate out the original ingredients?
4. If you were to remove the sawdust again, what would you do to make it easier?
5. How have you seen filtration used in real life? Explain.

Day Two
1. What do you notice about the plaster after the mixture sat for awhile? Why do you think this has happened? How did it make it easier to separate out the plaster? What is this process called?
2. What other situations have you ever noticed where settling has occurred?
3. Think of different ways that could be used to separate the sugar from the water. Which way seems to be the best one to you? Why?
4. What do you think could be done to speed up the evaporation process?
5. How did you determine how much water had evaporated? Explain your thinking process.
6. What was the most difficult part of separating the whole mixture? What could be done to make it easier?
7. Looking back at what you have experienced, how would you separate this mixture in a different way another time? Think about and explain ways that might make it work faster, or easier, or separate out the ingredients more completely.
8. What are some real-world situations in which mixtures of substances need to be separated? How

are filtration, settling, evaporation, or distillation used in these situations? (Think about home, school, the cafeteria, business, factories, the environment....) What modern technology do you know about that is used to make separation of mixtures more efficient?

Evidence of Learning

Ask the students to explain how they would separate a mixture of salt, sand, water, and beans. Have them respond in their science logs using pictures and words.

* Reprinted with permission from *Principles and Standards for School Mathematics*, 2000 by the National Council of Teachers of Mathematics. All rights reserved.

Messing With Mixtures

Key Question

How can the ingredients of a mixture be separated?

LEARNING GOALS

STUDENTS WILL:

1. identify what a mixture is; and
2. explore how filtration, settling, and evaporation can be used to separate a mixture.

Find the mass of: mixture and cup _____

−   empty cup _____
_____
                              mixture _____

mixture

empty
cup

drain well

sawdust

Find the mass of: sawdust and cup _____

−   empty cup _____
_____
                              sawdust _____

Describe ideas about how to separate the remaining ingredients.

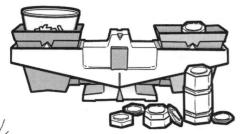

MESSING WITH MIXTURES

Please
Do Not
Disturb

sawdust

mixture

Messing With Mixtures

Day 2

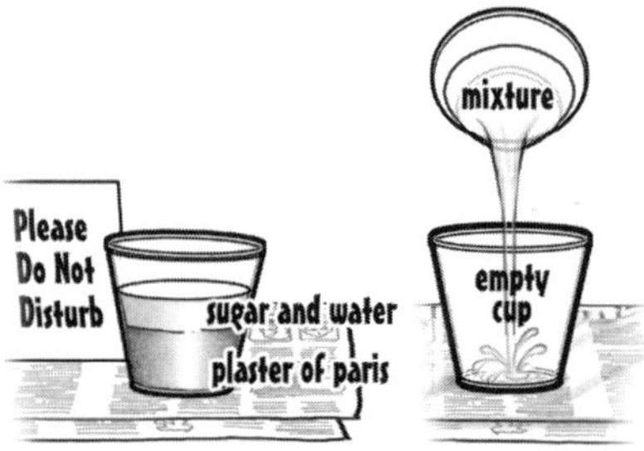

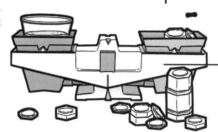

Find the mass of:

plaster of paris and cup ____________

− empty cup ____________

plaster of paris ____________

Describe ideas about how to separate the remaining ingredients.

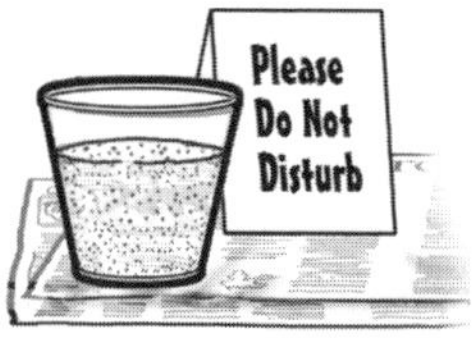

How long do you think it is going to take for the water to evaporate?

What do you think is going to be left in the cup after the water is gone?

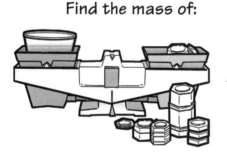

Find the mass of:

sugar and cup ____________

− empty cup ____________

sugar ____________

plaster of paris ____________

sawdust ____________

water ____________

+ sugar ____________

total ____________

Does the total mass of the component mixtures equal the mass you had at the beginning of the activity? Explain.

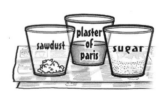

Connecting Learning

Day One

1. What do you observe about this mixture? How do you know that it is a mixture and not a compound?

2. What different ways can you think of to separate this mixture back into its original ingredients? Which idea do you think will work the best? Why do you think so?

3. What tools can you think of that might help you and your team separate out the original ingredients?

4. If you were to remove the sawdust again, what would you do to make it easier?

5. How have you seen filtration used in real life? Explain.

Day Two

1. What do you notice about the plaster after the mixture sat for awhile? Why do you think this has happened? How did it make it easier to separate out the plaster? What is this process called?

2. What other situations have you noticed where settling has occurred?

3. Think of different ways that could be used to separate the sugar from the water. Which way seems to be the best one to you? Why?

Connecting Learning

4. What could be done to speed up the evaporation process?

5. How did you determine how much water had evaporated? Explain.

6. What was the most difficult part of separating the whole mixture? What could be done to make it easier?

7. Looking back at what you have experienced, how would you separate this mixture in a different way another time? Think about and explain ways that might make it work faster, or easier, or separate out the ingredients more completely.

8. What are some real-world situations in which mixtures of substances need to be separated? How is filtration, settling, evaporation, or distillation used in these situations? (Think about home, school, the cafeteria, business, factories, the environment....) What modern technology do you know about that is used to make separation of mixtures more efficient?

9. What are you wondering now?

Topic
Properties of Elements

Key Question
How can you tell if a substance can be classified as a metal or non-metal?

Learning Goals
Students will:
1. identify that every element is classified as a metal, non-metal, or metalloid based on its individual properties; and
2. classify an element as a metal, non-metal, or metalloid based on observations of physical properties.

Guiding Documents
Project 2061 Benchmark
• *There are groups of elements that have similar properties, including highly reactive metals, less-reactive metals, highly reactive nonmetals (such as chlorine, fluorine, and oxygen), and some almost completely nonreactive gases (such as helium and neon). An especially important kind of reaction between substances involves combination of oxygen with something else—as in burning or rusting. Some elements don't fit into any of the categories; among them are carbon and hydrogen, essential elements of living matter.*

NRC Standards
• *Substances react chemically in characteristic ways with other substances to form, new substances (compounds) with different characteristic properties. In chemical reactions, the total mass is conserved. Substances often are placed in categories or groups if they react in similar ways; metals is an example of such a group.*
• *Think critically and logically to make the relationships between evidence and explanations.*

Science
Physical science
   properties of elements

Integrated Processes
Observing
Comparing and contrasting
Communicating
Collecting and recording data

Interpreting data
Inferring
Generalizing

Materials
For each student group:
   one small paper clip
   one piece of copper wire, 6 cm long
   one fishing weight
   one piece of charcoal (see *Management 1*)
   one aluminum pop tab
   one piece of graphite (see *Management 2*)
   one piece of mylar, 6 cm square
   one circuit tester (see *Management 3*)

For the class:
   a hammer (see *Management 4*)
   a collection of newspapers
   Properties of Metals transparency

Background Information
   Metals have characteristics that make them different from non-metals. Metals have luster, which means they easily reflect light. Metals also are malleable. This means that they can be hammered or pressed without breaking. Metals are conductors of electricity and heat. Metals react with acids, although we will not explore this characteristic with this experience. Non-metals, in very simple terms, do not share the same characteristics as do metals. The other group of elements is called metalloids. Metalloids have some properties of metals as well as non-metals.

Management
1. Use charcoal briquettes that do not have the lighter fluid added to them. Break a briquette into small pieces.
2. Graphite bars can be purchased in craft or art stores.
3. Construct a conductivity meter from a 9-volt battery, a bulb and bulb holder, and three pieces of insulated wire. Make sure the students test whether a material conducts electricity by placing the material between the two wires.

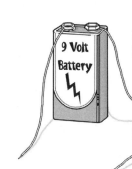

4. Make a transparency of the page *Properties of Metals*.

5. The students will test for malleability by placing the material between newspaper and hitting it five times with a hammer. They will need five thicknesses of paper between the hammer and the test objects.

Procedure

1. Ask the *Key Question* and state the *Learning Goals*.

2. Show the transparency *Properties of Metals*. Discuss with the students the background information about metals, non-metals, and metalloids.

3. Distribute the student recording sheet and the materials.

4. Point out the station that they will come to test for malleability. Stress the importance of safety when using the hammer.

5. Circulate as the students are working with the materials.

6. Have each group share their findings and their classification as to if the materials are metals, non-metals or metalloids.

Connecting Learning

1. Which items were metals? [paper clip, copper wire, fishing weight, and aluminum pop tab] How do you know?

2. What observations about graphite would make you think it was a metal? What observations would make you think it was a non-metal?

3. Why is it important to conduct multiple tests when you are identifying materials?

4. Why do you think we have a group of elements called metalloids?

5. What are you wondering now?

Extension

Use the periodic table to show the location of the metals, non-metals, and metalloids. If possible, have students shade the elements of the different classifications with different colors.

9 Volt Battery

Messing With Metals

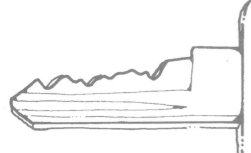

•Key Question•
How can you tell if a substance can be classified as a metal or non-metal?

Learning Goals
The students will:

1. identify that every element is classified as a metal, non-metal, or metalloid based on its individual properties; and

2. classify an element as a metal, non-metal, or metalloid based on observations of physical properties.

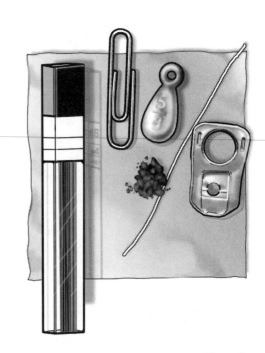

Properties
of Metals

Metals have luster

Metals are malleable

Metals are good
conductors of electricity

Metals react with acids

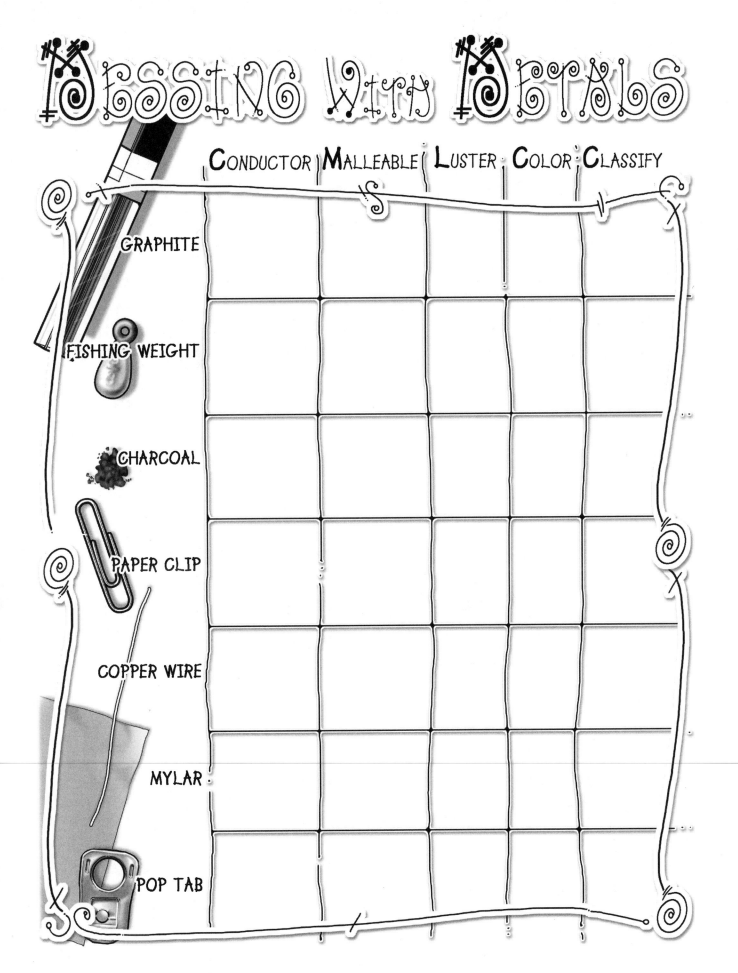

MESSING WITH METALS

	Conductor	Malleable	Luster	Color	Classify
GRAPHITE					
FISHING WEIGHT					
CHARCOAL					
PAPER CLIP					
COPPER WIRE					
MYLAR					
POP TAB					

Messing With Metals

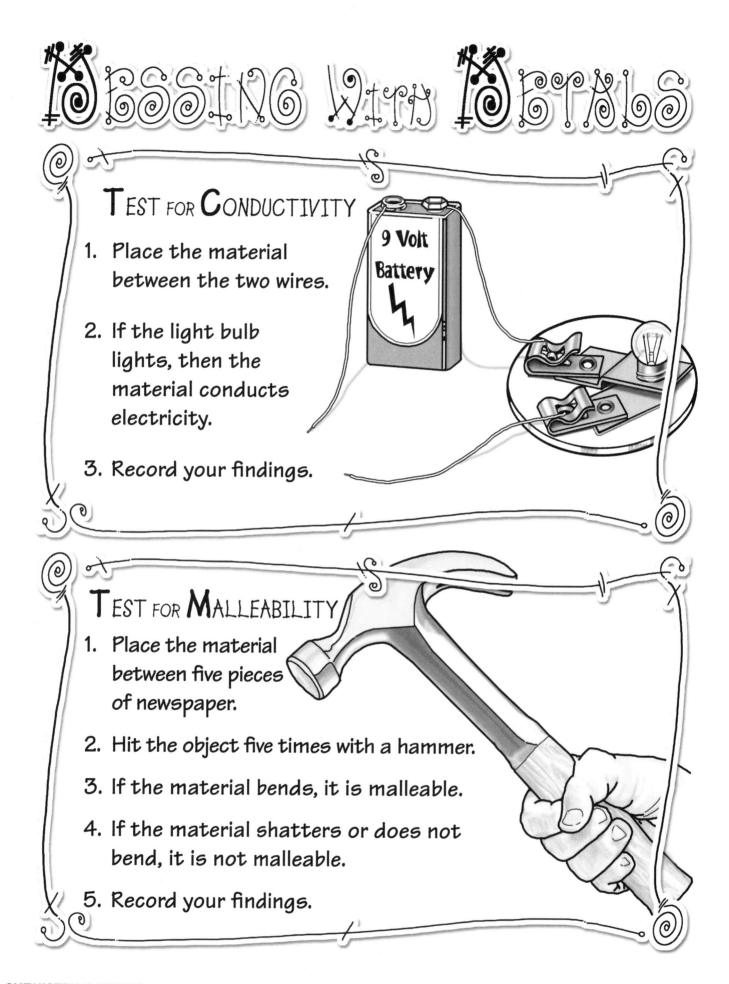

Test for Conductivity

1. Place the material between the two wires.

2. If the light bulb lights, then the material conducts electricity.

3. Record your findings.

9 Volt Battery

Test for Malleability

1. Place the material between five pieces of newspaper.

2. Hit the object five times with a hammer.

3. If the material bends, it is malleable.

4. If the material shatters or does not bend, it is not malleable.

5. Record your findings.

Connecting Learning

1. Which items were metals? How do you know?

2. What observations about graphite would make you think it was a metal? What observations would make you think it was a non-metal?

3. Why is it important to conduct multiple tests when you are identifying materials?

4. Why do you think we have a group of elements called metalloids?

5. What are you wondering now?

Topic
Solutions, Solvents, and Solutes

Key Question
Which solvent will hold the most solute in solution?

Learning Goals
Students will:
1. distinguish between a solution, solvent and solute; and
2. determine the limit of each solvent with each solute.

Guiding Documents
Project 2061 Benchmark
- *Scientific investigations may take many different forms, including observing what things are like or what is happening somewhere, collecting specimens for analysis, and doing experiments. Investigations can focus on physical, biological, and social questions.*

NRC Standard
- *A substance has characteristic properties, such as density, a boiling point, and solubility, all of which are independent of the amount of the sample. A mixture of substances often can be separated into the original substances using one or more of the characteristic properties.*

NCTM Standards 2000*
- *Systematically collect, organize, and describe data*
- *Construct, read, and interpret tables, charts, and graphs*
- *Make inferences and convincing arguments that are based on data analysis*

Math
Measurement
    volume

Science
Physical science
    matter

Integrated Processes
Observing
Inferring
Comparing and contrasting
Collecting and recording data
Interpreting data

Material
For each group:
    300 mL of rubbing alcohol
    300 mL of water
    100 mL of kosher salt
    100 mL of alum
    100 mL of Epsom salt
    250 mL beaker (see *Management 1*)
    100 mL graduated cylinder
    2 plastic teaspoons

Background Information
Dissolving is a physical change in which the molecules, atoms, or ions of a substance dissociate from that substance and mix with the molecules of the dissolving liquid. The original substance is called the *solute* and the dissolving liquid is called the *solvent*. The new mixture is called a *solution*. All solvents have a limit as to how much of a given substance they can hold in solution. Once that limit is reached, no more solute will be able to be dissolved in the solvent. Although solutions, solvents, and solutes are evident in the students' everyday life, these ideas do not seem important unless they can be brought to the students as something they can experience.

Management
1. Any flat bottom container will work as long as it holds 250 mL of liquid.
2. Make sure the students clean out the beaker between each solvent and solute combination.
3. Caution the students not to taste any of the solvents or solutes.

Procedure
1. Ask the *Key Question* and state the *Learning Goals*.
2. Distribute the materials to the students. Tell them that they will begin with the kosher salt and water. Point out that the kosher salt is the solute and the water is the solvent.
3. Direct the students to pour 100 mL of water into the beaker.
4. Show the students how to add a level teaspoon of salt into the water and stir with the second spoon.
5. Point out to the students that the crystals of salt dissolve. Tell the students this is now a solution of kosher salt and water. Direct them to continue adding a teaspoon of the kosher salt at a time and stir until no more can be dissolved in the water.

6. Tell the students to record the number of teaspoons of salt that could be dissolved into 100 mL of water.
7. Repeat steps 3 through 6 for the alum and then the Epsom salt.
8. Repeat steps 3 through 6 for the kosher salt, alum, and Epsom salt with the rubbing alcohol as the solvent.

Connecting Learning
1. In your own words, describe a solute, a solvent, and a solution?
2. Of the three—solute, solvent, and solution—which was the salt?
3. In order for the salt, a solute, to become a solution, what did you have to do? [dissolve it in water]
4. Which solvent seemed to be able to hold the most solute?
5. Which solute was the easiest to dissolve in the solvent?
6. Water is called the universal solvent. Based on what you discovered in this investigation, why do you think scientists refer to water like this?
7. In general terms, the more atoms in a molecule the less that that molecule will be able to be held in a solution. Each of the three solutes we used is composed of different combinations and numbers of atoms. Based on your data, which solute has the most number of atoms?
8. How is a solution related to a solvent and solute? [When solvents and solutes mix, they make a solution.]
9. What solvents, solutes, and solutions are a part of your everyday life? [sugar dissolved in drinks is one example]
10. What are you wondering now?

Evidence of Learning
1. Listen for student talk during the *Connecting Learning* questions.
2. Check for accuracy in filling out the student sheets and graphs.

* Reprinted with permission from *Principles and Standards for School Mathematics*, 2000 by the National Council of Teachers of Mathematics. All rights reserved.

Involving Dissolving

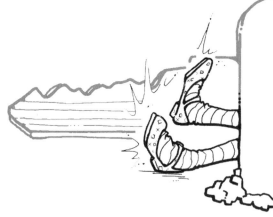

Key Question

Which solvent will hold the most solute in solution?

Learning Goals

Students will:

1. distinguish between a solution, solvent, and solute; and
2. determine the limit of each solvent with each solute.

Involving Dissolving

How much solute will each solvent hold?

1. Use your own words to define a solution, a solvent, and solute.

	100 mL water	100 mL rubbing alcohol
kosher salt	tsp	tsp
alum	tsp	tsp
Epsom salt	tsp	tsp

2. Describe how salt is a solute in this investigation.

3. Which solvent seemed to be able to hold the most solute?

4. Water is called the universal solvent. From what you learned in this investigation, explain why this is true.

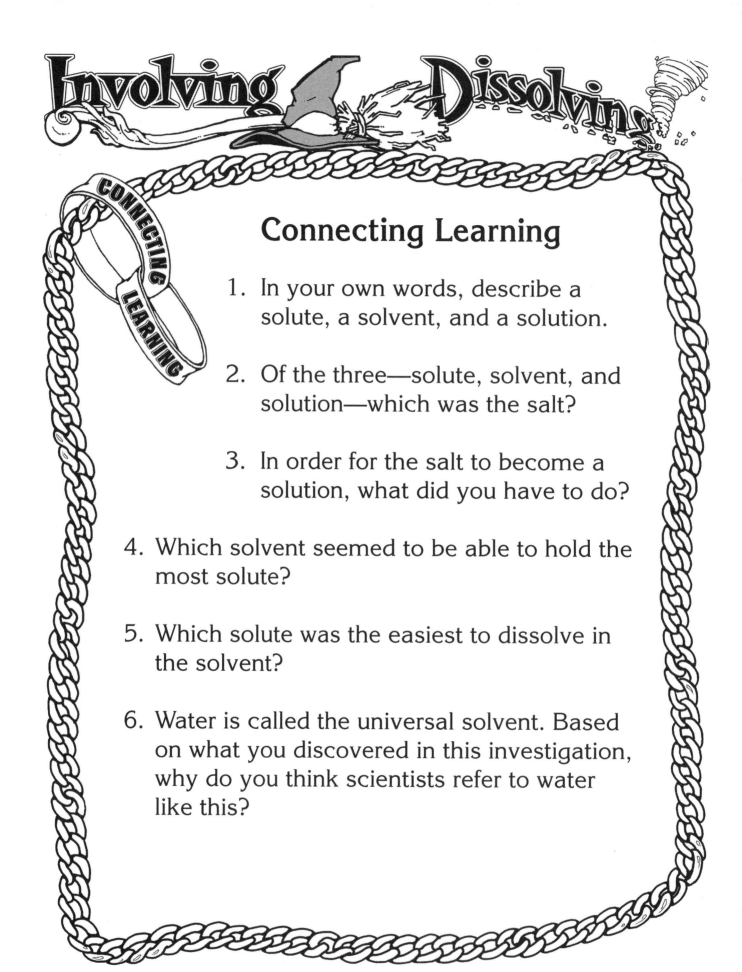

Involving Dissolving

Connecting Learning

1. In your own words, describe a solute, a solvent, and a solution.

2. Of the three—solute, solvent, and solution—which was the salt?

3. In order for the salt to become a solution, what did you have to do?

4. Which solvent seemed to be able to hold the most solute?

5. Which solute was the easiest to dissolve in the solvent?

6. Water is called the universal solvent. Based on what you discovered in this investigation, why do you think scientists refer to water like this?

69

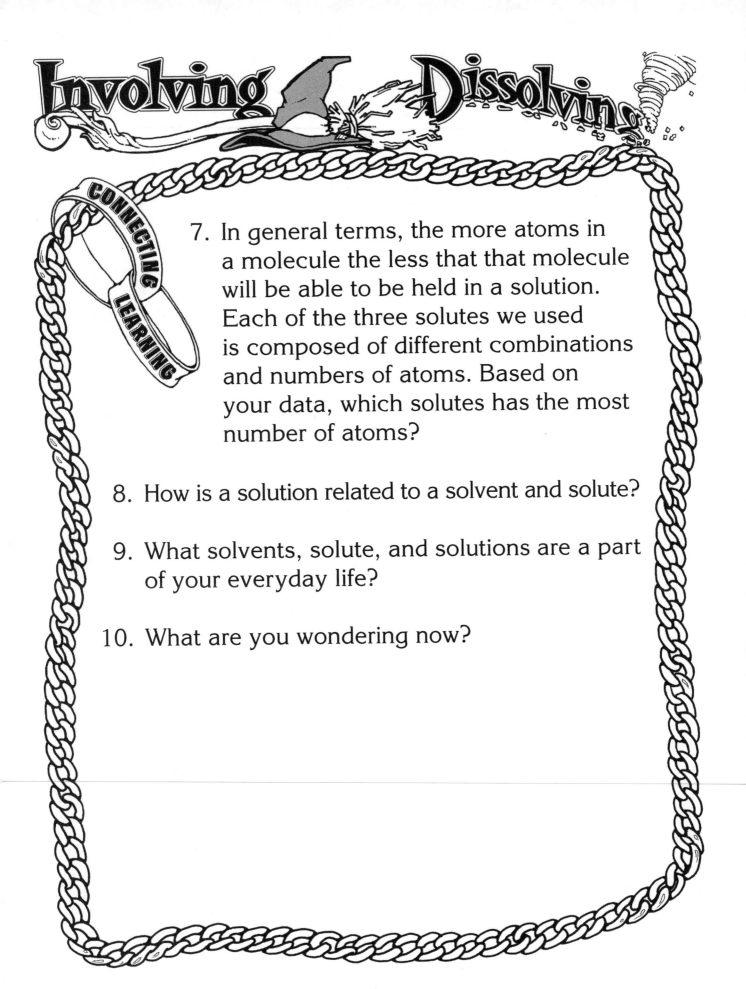

7. In general terms, the more atoms in a molecule the less that that molecule will be able to be held in a solution. Each of the three solutes we used is composed of different combinations and numbers of atoms. Based on your data, which solutes has the most number of atoms?

8. How is a solution related to a solvent and solute?

9. What solvents, solute, and solutions are a part of your everyday life?

10. What are you wondering now?

Watch it... BURN

Topic
Changes in States of Matter

Key Question
What states of matter can you observe as you watch a candle burn?

Learning Goals
Students will:
1. identify properties of matter,
2. identify three states or phases while a candle is burning, and
3. observe matter changing states.

Guiding Documents
Project 2061 Benchmarks
- *Heating and cooling cause changes in the properties of materials. Many kinds of changes occur faster under hotter conditions.*
- *Things that give off light often also give off heat. Heat is produced by mechanical and electrical machines, and any time one thing rubs against something else.*

NRC Standards
- *Different kinds of questions suggest different kinds of scientific investigations. Some investigations involve observing and describing objects, organisms, or events; some involve collecting specimens; some involve experiments; some involve seeking more information; some involve discovery of new objects and phenomena; and some involve making models.*
- *Materials can exist in different states—solid, liquid, and gas. Some common materials, such as water, can be changed from one state to another by heating or cooling.*
- *Objects have many observable properties, including size, weight, shape, color, temperature, and the ability to react with other substances. Those properties can be measured using tools, such as rulers, balances, and thermometers.*

NCTM Standard 2000*
- *Understand such attributes as length, area, weight, volume, and size of angle and select the appropriate type of unit for measuring each attribute*

Math
Measurement
   length
   time
   mass

Science
Physical science
   matter

Integrated Processes
Observing
Comparing and contrasting
Classifying
Communicating
Predicting
Collecting and recording data
Interpreting data

Materials
For the teacher:
   matches or lighter
   group observation charts (see *Management 4* and *5*)
   fire extinguisher or water source

For each student group:
   balance
   masses
   birthday candle
   clay
   aluminum foil square, 6 cm x 6 cm
   colored pencils
   metric rulers (see *Management 6*)
   mini-book (see *Management 8*)
   optional: cardboard square, 10.5 cm x 10.5 cm
     (see *Management 7*)

Background Information
Matter can undergo both physical and chemical changes. A physical change involves changing the form of an object or the state of the object. If you break a wooden match, it undergoes a physical change. Substances that undergo a chemical change are turned into new substances that are different chemically than the originals. A wooden match that has been burned has undergone a chemical change because the wood no longer has the same chemical properties it did before it was burned.

Students need to begin their study of chemistry by learning how to make careful observations as well as accurate measurements. An important outcome as students initially are engaged in chemistry activities is the ability of the student to recognize physical and chemical changes in matter. Conceptually, chemical changes are more difficult for the students to recognize and understand. Sustained exposure to the basic concepts of chemistry will provide students with the conceptual tools for identifying and describing chemical changes.

This activity utilizes a birthday candle. Most children have seen a birthday candle but few have ever really observed one. Careful observation of a burning candle creates opportunities for students to identify changes occurring in the candle and to determine which of the observed changes they can measure (length and mass). Although the students can observe that the flame is hot, they do not have a tool with which they can measure the temperature of the candle flame. They will need to rely upon measurements made by scientists who have the techniques and tools required to measure the various temperatures that occur in a candle flame.

Management

1. Organize the students into groups of three.
2. If necessary, review with the students how to read a ruler to the nearest mm and how properly to use a balance.
3. Prepare 2 three-vertical column charts with chart paper or bulletin board paper. Place the pictures of the candle, clay, and aluminum foil in the columns on both charts. Label one chart *Physical Properties* and one *Measurement*.

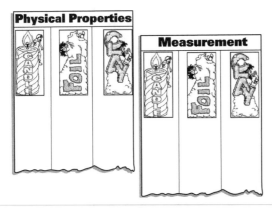

4. Prepare a third chart with the drawing of the burning candle label. Divide this chart horizontally with two columns. Label the top section *Observed Changes* and the bottom section *Measurements*.
5. Copy a sheet of the rulers for the students to use for the observation of the

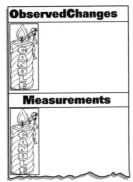

candles. These mini rulers will permit measuring the circumference as well as length of the objects.
6. If you have primer balances, cut a piece of cardboard, 10.5 cm by 10.5 cm square, to place in the top of the balance.

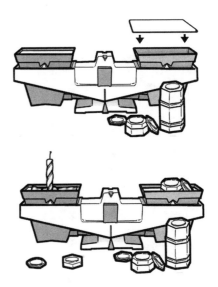

7. Prepare enough mini-books of *The Candle Flame Exposed* as you will need. One book for every two students should be sufficient.
8. Have a fire extinguisher or water source available for safety purposes.

Procedure

Part One

1. Ask the students the *Key Question* and state the *Learning Goals*.
2. Distribute the materials to the student groups.
3. Tell the students they will be making observations of the physical properties of the three items they were given. The students will need to record the data in the observation section of the student sheet. Encourage them to make as many observations as they are able.
4. Direct the students in a discussion on what measurements can be made on the foil, candle, and clay. Tell them to record the measurements in the measurement section of the student page. Remind them to be sure to include the unit of measure.
5. Discuss and record on the group charts observations of physical properties as well as measurements.

Part Two

1. Review the *Key Question* and the *Learning Goals*.
2. Distribute the *Open Flame Safety Rules*. Review safety procedures and let the students read and sign the *Open Flame Safety Rules*.
3. Instruct the students to form a hemisphere with the clay to serve as a candle holder. Have them put the clay in the center of the aluminum foil square and stand the candle in the clay.

4. Tell the students to place the piece of aluminum foil with the clay and candle on one side of the balance (on top of the cardboard piece if you are using primer balances). Direct them to add masses to the other side until the pans of the balance are equalized. Tell students that they are to watch what happens to the balance as the candle burns.

5. Ask the students to predict some of the changes they expect to see. Tell one student in each group to record the height of the candle as it burns. Caution them to keep the ruler away from the flame. Encourage them to record the height to the nearest millimeter every 30 seconds. As a class define the height of the candle as the height of the wax.

6. Light each group's candle and encourage students to make as many observations as they are able to as the candle burns. Make sure they record the time the candle is lit and the time when they extinguish the candle. Direct the student to use the colored pencils to record the colors seen in the flame of the candle.

7. Tell the students to record any changes in the candle, clay, and foil.

8. Lead a class discussion and record the observations students made on the class charts. Tell the students to graph the data they collected on the burning of the candle.

9. Direct the students in reading and discussing *The Candle Flame Exposed* mini-book.

Connecting Learning

1. What states of matter were you able to observe in this investigation? [solid, liquid, and gas]

2. Why do you think we burned the candle on the balance? [The balance showed that the candle's mass changed during the burning, indicating that a chemical change was taking place. The solid candle was being converted into heat and gas as it burned.]

3. What did the graph show you about how a birthday candle burns? From the graph, could you predict the amount of time a 10-cm birthday candle would burn? Explain.

4. What caused the change of states in this investigation? [The burning of the candle caused the change.]

5. Why is it necessary to list the unit of measurement with observations? [The unit describes the magnitude of measure. For example, 1 mm is small compared to 1 km.]

6. How do you think you would make the clay and aluminum foil move from a solid state to a liquid state? [If you applied enough heat, you would be able to do it.]

Evidence of Learning

1. Look for descriptions in the observations as you record data on the group chart.

2. Look for accurate recording for units of measurement.

* Reprinted with permission from *Principles and Standards for School Mathematics*, 2000 by the National Council of Teachers of Mathematics. All rights reserved.

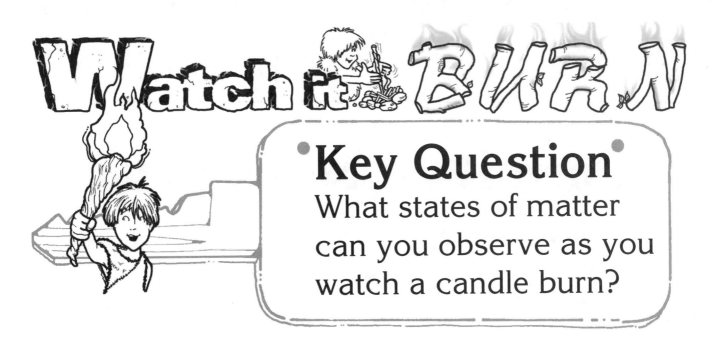

Watch it BURN

°Key Question°
What states of matter can you observe as you watch a candle burn?

Learning Goals

Students will:

1. identify properties of matter,
2. identify three states (phases) of matter while a candle is burning, and
3. observe matter changing states.

Open Flame Safety Rules

1. Listen to or read instructions carefully before attempting to do anything.

2. Tie back long hair.

3. Roll up loose sleeves.

4. Never reach across an open flame.

5. Keep the area around the open flame clear.

6. Tell your teacher if any spills or accidents occur.

7. Know the location of the fire extinguisher and how to use it.

Group Members' Signatures

_____

_____

_____

Teacher's Signature

_____

Watch it BURN

millimeter strips

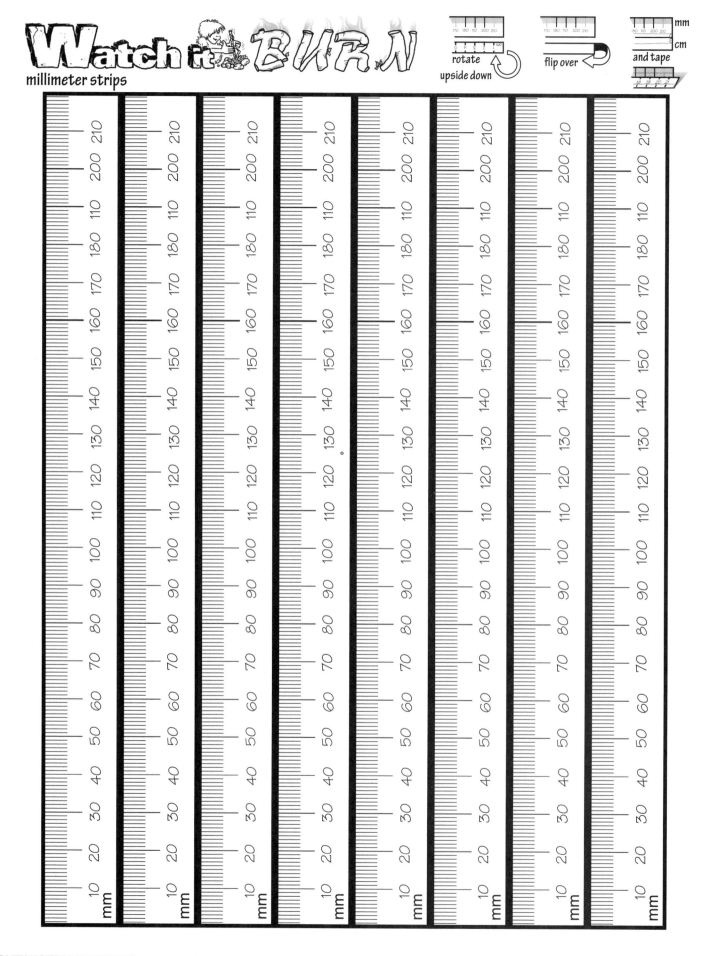

rotate upside down   flip over   and tape

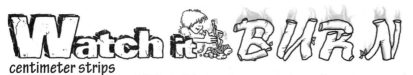

Watch it BURN

centimeter strips

rotate
upside down
flip over
mm
cm
and tape

cm
1
2
3
4
5
6
7
8
9
10
11
12
13
14
15
16
17
18
19
20
21

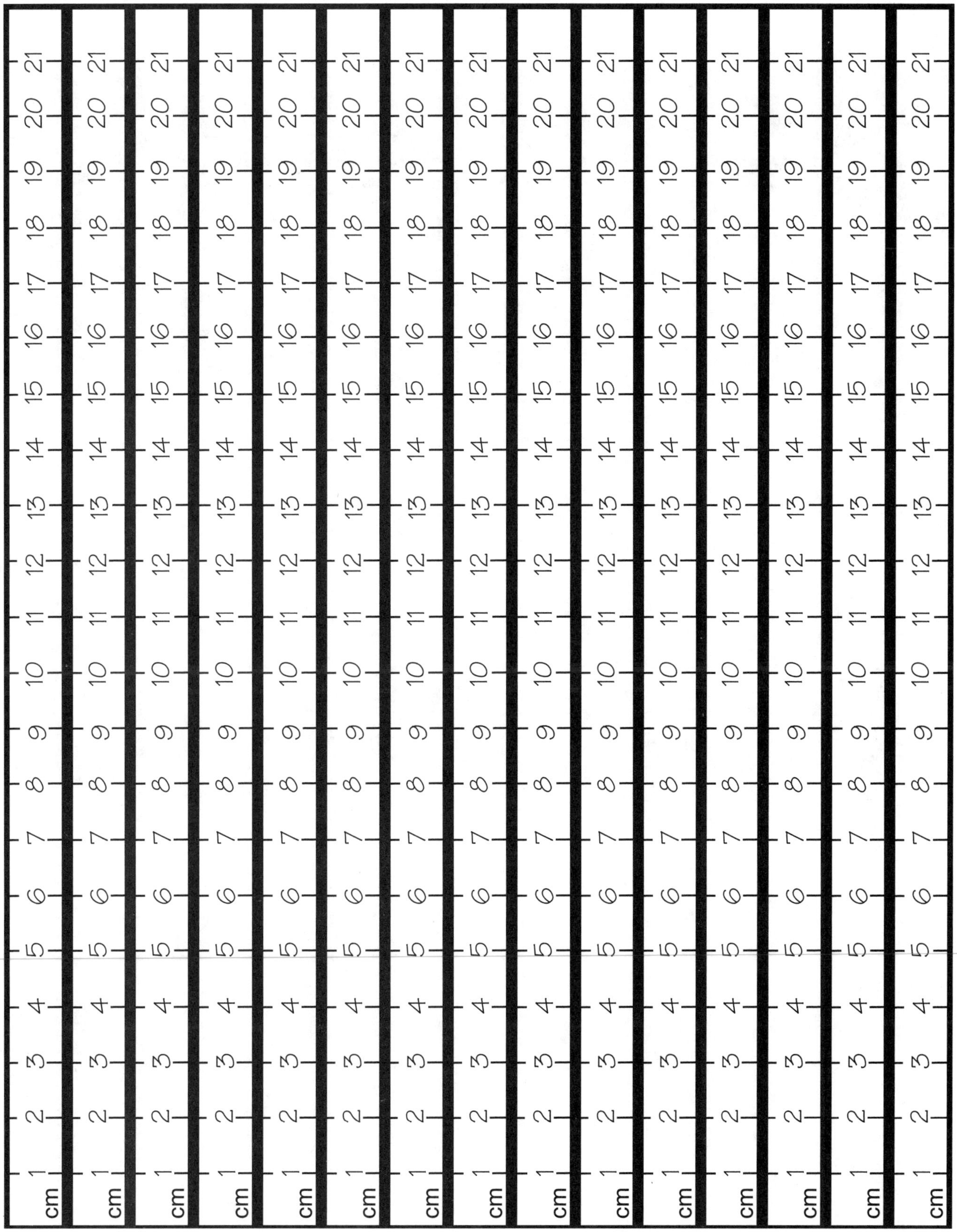

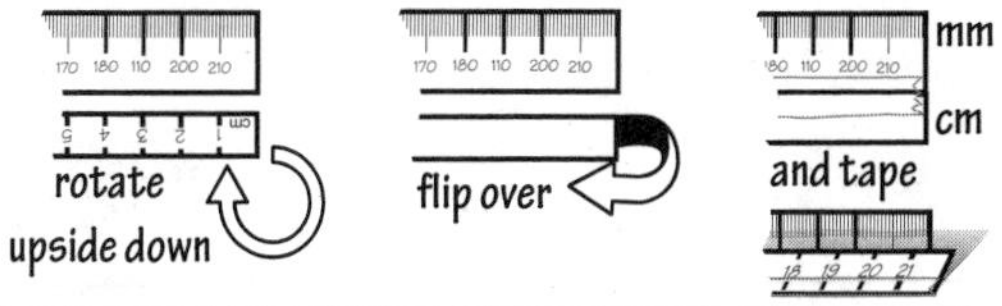

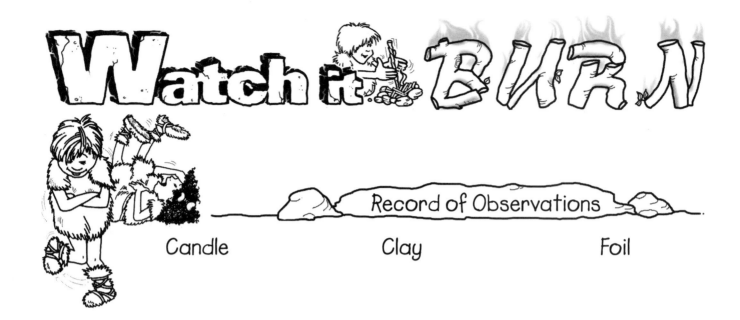

Watch it... BURN

Record of Observations

Candle   Clay   Foil

Record of Measurements

Candle   Clay   Foil

Watch it... BURN

Time candle was lit: _____

Time candle was extinguished: _____

Observations of changes:

Drawing of candle flame

Measurements:

Initial Measurement in mm _____
Record the height of the candle every 30 seconds.

30 sec. _____	4.5 min. _____
1 min. _____	5 min. _____
1.5 min. _____	5.5 min. _____
2 min. _____	6 min. _____
2.5 min. _____	6.5 min. _____
3 min. _____	7 min. _____
3.5 min. _____	7.5 min. _____
4 min. _____	8 min. _____

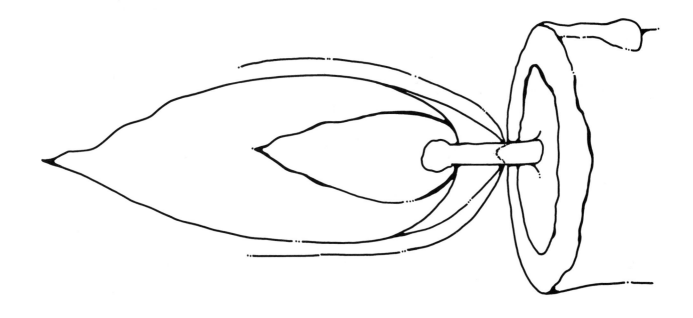

cut

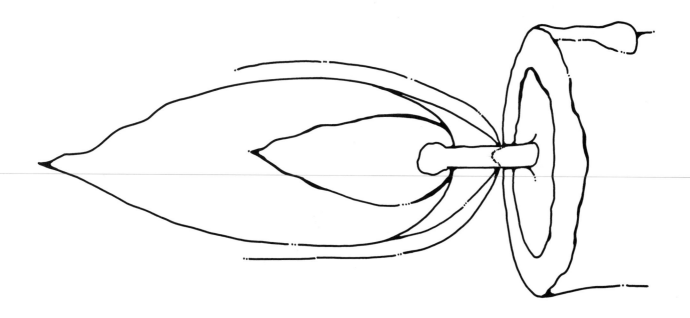

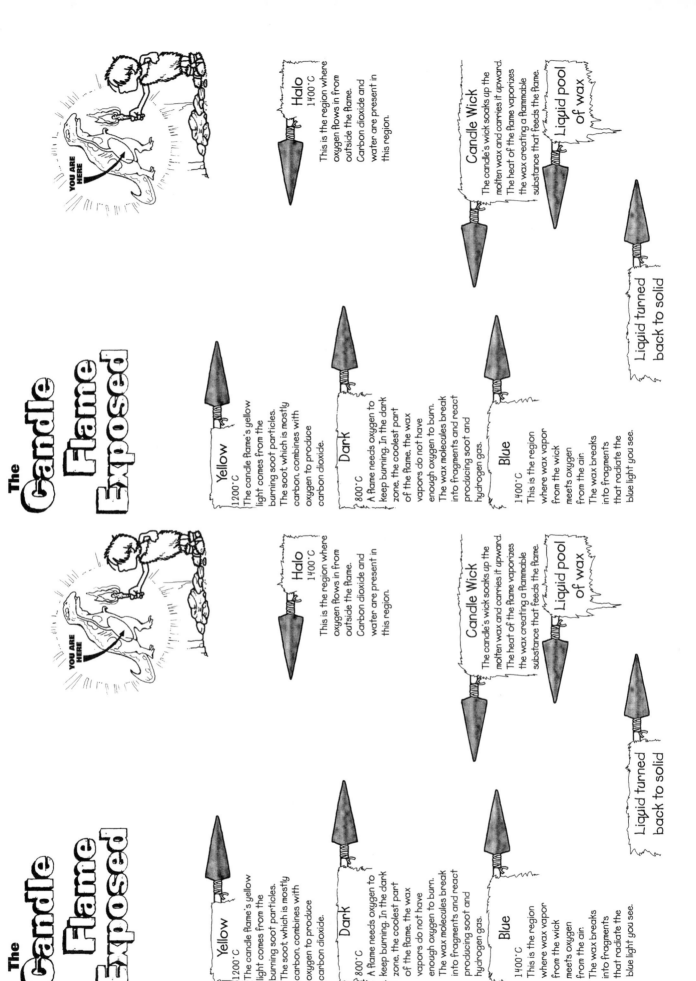

The Candle Flame Exposed

Yellow
1200°C
The candle flame's yellow light comes from the burning soot particles. The soot, which is mostly carbon, combines with oxygen to produce carbon dioxide.

Dark
800°C
A flame needs oxygen to keep burning. In the dark zone, the coolest part of the flame, the wax vapors do not have enough oxygen to burn. The wax molecules break into fragments and react producing soot and hydrogen gas.

Blue
1400°C
This is the region where wax vapor from the wick meets oxygen from the air. The wax breaks into fragments that radiate the blue light you see.

Halo
1400°C
This is the region where oxygen flows in from outside the flame. Carbon dioxide and water are present in this region.

Candle Wick
The candle's wick soaks up the molten wax and carries it upward. The heat of the flame vaporizes the wax creating a flammable substance that feeds the flame.

Liquid pool of wax

Liquid turned back to solid

YOU ARE HERE

The Candle Flame Exposed

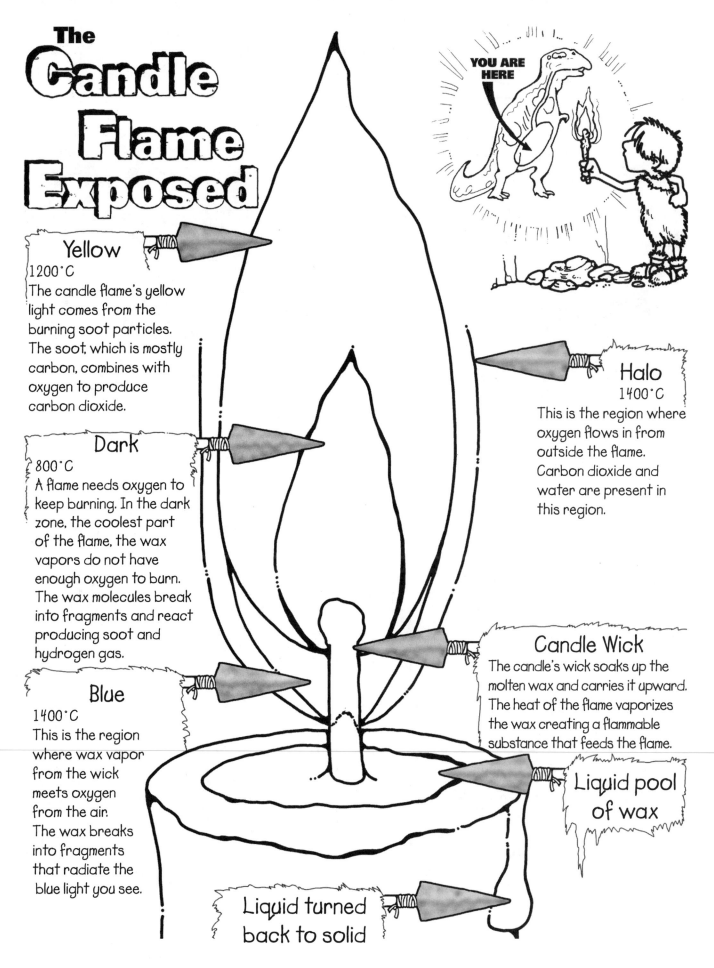

YOU ARE HERE

Yellow
1200°C
The candle flame's yellow light comes from the burning soot particles. The soot, which is mostly carbon, combines with oxygen to produce carbon dioxide.

Dark
800°C
A flame needs oxygen to keep burning. In the dark zone, the coolest part of the flame, the wax vapors do not have enough oxygen to burn. The wax molecules break into fragments and react producing soot and hydrogen gas.

Blue
1400°C
This is the region where wax vapor from the wick meets oxygen from the air. The wax breaks into fragments that radiate the blue light you see.

Halo
1400°C
This is the region where oxygen flows in from outside the flame. Carbon dioxide and water are present in this region.

Candle Wick
The candle's wick soaks up the molten wax and carries it upward. The heat of the flame vaporizes the wax creating a flammable substance that feeds the flame.

Liquid pool of wax

Liquid turned back to solid

Watch it BURN

Make a graph of your data. Use the graph to predict the amount of time a 10-cm birthday candle will burn. _____

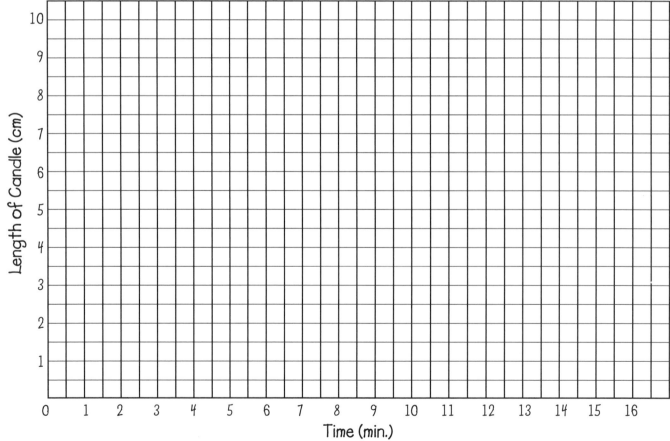

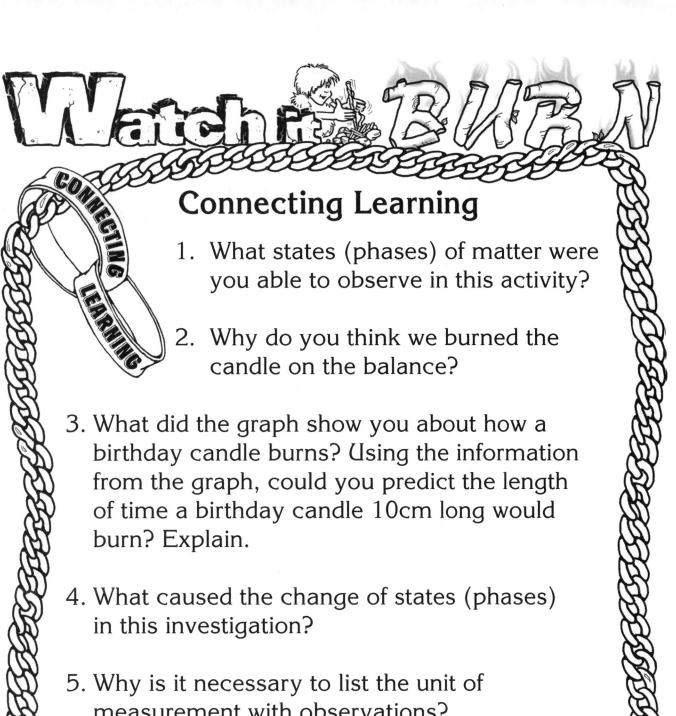

Connecting Learning

1. What states (phases) of matter were you able to observe in this activity?

2. Why do you think we burned the candle on the balance?

3. What did the graph show you about how a birthday candle burns? Using the information from the graph, could you predict the length of time a birthday candle 10cm long would burn? Explain.

4. What caused the change of states (phases) in this investigation?

5. Why is it necessary to list the unit of measurement with observations?

6. How do you think you would make the clay and aluminum foil move from a solid state to a liquid state?

7. What are you wondering now?

Kool Kups

Topic
States of Matter

Key Question
How can you observe a change in state by observing a glass of flavored fruit drink?

Learning Goals
Students will:
1. observe a change in state in water;
2. identify the three states of matter; and
3. be able to explain, using evidence from the investigation, that the water came from the air not the glass.

Guiding Documents
Project 2061 Benchmarks
- *Scientists do not pay much attention to claims about how something they know about works unless the claims are backed up with evidence that can be confirmed and with a logical argument.*
- *Heating and cooling cause changes in the properties of materials. Many kinds of changes occur faster under hotter conditions.*

NRC Standards
- *Think critically and logically to make the relationships between evidence and explanations*
- *Materials can exist in different states—solid, liquid, and gas. Heating or cooling can change some common materials, such as water, from one state to another.*

NCTM Standard 2000*
- *Propose and justify conclusions and predictions that are based on data and design studies to further investigate the conclusions or predictions*

Math
Data analysis

Science
Physical science
    matter

Integrated Processes
Observing
Relating
Communicating
Interpreting data
Inferring

Materials
For each group:
    ice (see *Management 2*)
    2 clear plastic cups, 8 oz.
    flavored drink mix (see *Management 3*)
    2 pieces of white paper towels
    2 thermometers
    4 rubber bands

Background Information
The air around us is made up of different types of gases. One of the gases in air is called water vapor.

Energy transfer is needed for a phase change in matter. When water is heated, water moves from a liquid state to a gaseous state. When water vapor is cooled, it moves from a gas back to a liquid. The air surrounding the cup in this investigation is cooled to the point that the water vapor in the air surrounding the cup moves from a gas back to a liquid state.

The flavored drink mix is important in the activity so that the students will have evidence that the water on the outside of the cup did not come through the walls of the cup or it would be the same color and flavor as the flavored drink mix.

Management
1. This activity works best in pairs so that the students can drink the flavored drink mix at the end of the investigation. The students can fair share the ice and drink.
2. Be sure to have sufficient ice so that each student group will have a cup of ice.
3. Prepare enough drink mix so that you will be able to fill two cups for each student group. Select a red or orange colored drink so that the students will be able to clearly see the color.
4. Attach the thermometers to the outside on the cups with the rubber bands so that the scale can still be read.

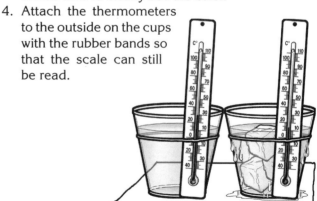

Procedure

1. Ask the *Key Question* and state the *Learning Goals*.
2. Distribute the materials to each student group.
3. Direct the students to place a paper towel on their work surfaces and put the cups on the paper towel. Show the students how to attach the thermometers to the outside of the cups. Read and record the temperature of each cup on the student sheet. Tell the students to read and record the temperatures of both cups on the temperature graph on the student page every three minutes.
4. Tell the students to fill one cup with ice cubes. Do not add any ice to the other cup.
5. Direct the student to fill each cup with the flavored drink.
6. After four or five minutes, ask the students, "Do you notice anything forming on the outside of either cup? What does it look like?" Remind the students to continue recording the temperature of the cups on the student page every three minutes.
7. Ask the students to touch the outside of the cups with their fingers. Is there any difference between the way the cups feel? Does either of the cups feel wet? [The cup with the ice should feel wet and should feel cooler. If it doesn't, wait a few minutes and try again.]
8. Have the students wipe the outside of the cup that does not have the ice in it with the white paper towel. Ask the students, "What do you observe?" Tell the students to wipe the outside of the cup with the ice in it with the paper towel. Ask the students, "What do you observe?" [The students should see a wet spot.] Ask the students, "Is the wet spot the same color as the flavored drink?" Ask the students, "How does this help us see that the water did not come from the cup?" [The spot would be the same color as the drink mix if it came through the side of the cup.]
9. Ask the students "If the water did not come from the cup, where do you think it came from?"
10. Demonstrate how to place a hand close to the cup without touching it. Have each student place his or her hands close to each cup then ask, "What do you notice about the air around the cups? [The students should be able to feel the air near the cup with ice in it is cooler.]
11. Tell the students that the ice inside the cup has cooled the air around the cup. The water vapor in the air condensed (turned from water vapor to water) on the surface closest to the cooled air and that surface was the outside of the cup.
12. Lead a class discussion so that the students will be able to identify that the ice cubes in the cup are the solid form of water, they can see the liquid form of water on the outside of the cup. They now have evidence that the air was holding the water vapor.

13. Direct the students to label the three states of matter on the student record sheet.

Connecting Learning

1. What are the three common states of matter on Earth?
2. What evidence do we have that tells us the water came from the air and not the glass?
3. Why do you think we used the thermometers in this investigation? [The thermometers gave us numeric evidence of a change in temperature.]
4. When have you seen droplets of water on the surface of other objects? [The students should be able to relate this experience to morning dew.]
5. What do you think would happen if we used Styrofoam cups instead of plastic cups?
6. What are you wondering now?

Evidence of Learning

1. Listen for student explanation as they make observations and comments during the investigation.
2. Look for correct recording on the student's sheets.

Kool Kups

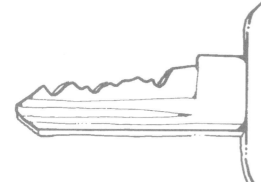

Key Question

How can you observe a change in state (phase) by observing a glass of flavored fruit drink?

Learning Goals

Students will:

1. observe a change in state (phase) in water;

2. identify the three states (phases) of matter; and

3. be able to explain, using evidence from the investigation, where the water on the outside of the cup came from.

 # Kool Kups

Complete the table.

	With Flavored Drink							
	start (C°)	3 min. (C°)	6 min. (C°)	9 min. (C°)	12 min. (C°)	15 min. (C°)	18 min. (C°)	21 min. (C°)
Cup One with ice								
Cup Two without ice								

Wipe the outside of cup one. What do you observe?

Wipe the outside of cup two. What do you observe?

What variable do you think caused the water to collect on the outside of the cup?

If the water did not come from the cup, where do you think it came from?

Label the three states of matter.

What do you notice about the air around the cups?

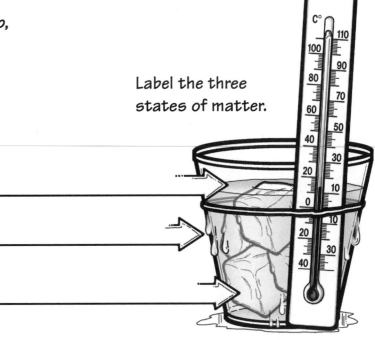

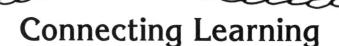

Connecting Learning

1. What are the three common states (phases) of matter on Earth?

2. What evidence do we have that tells us the water came from the air and not the glass?

3. Why do you think we used the thermometers in this investigation?

4. When have you seen droplets of water on the surface of other objects?

5. What do you think would happen if we used Styrofoam cups instead of plastic cups? Why?

6. What are you wondering now?

Product Testing

Topics
Chemical Change
Physical Properties of Matter

Key Question
What is the best formula to use to make Glubber?

Learning Goals
The students will:
1. identify and utilize percentage,
2. compare and contrast properties of matter, and
3. recommend a product formula based on experimental data.

Guiding Documents
Project 2061 Benchmark
- *When a new material is made by combining two or more materials, it has properties that are different from the original materials. For that reason, a lot of different materials can be made from a small number of basic kinds of materials.*

NRC Standard
- *Substances react chemically in characteristic ways with other substances to form new substances (compounds) with different characteristic properties. In chemical reactions, the total mass is conserved. Substances often are placed in categories or groups if they react in similar ways; metals is an example of such a group.*

NCTM Standards 2000*
- *Propose and justify conclusions and predictions that are based on data and design studies to further investigate the conclusions or predictions*
- *Develop, analyze, and explain methods for solving problems involving proportions, such as scaling and finding equivalent ratios*

Math
Data analysis
Measurement

Science
Physical science
    chemical change
    physical properties of matter

Integrated Processes
Observing
Comparing and contrasting

Inferring
Communicating
Collecting and recording data
Interpreting data
Drawing conclusions

Materials
For each group:
    white glue (see *Management 7*)
    100 mL graduated cylinders
    saturated borax solution (see *Management 3*)
    two clear plastic cups, 8-oz.
    two plastic spoons (see *Management 8*)
    black permanent marking pen
    three plastic zipper-type bags, pint size
    paper towels

For the Testing Stations:
    one meter stick
    three large paper clips
    task cards (see *Management 6*)
    one set of washable markers

Background Information
A chemical change is the interaction when substances combine to form a new substance. A new substance has its own set of properties. For example, when charcoal burns, the carbon atoms that make up charcoal combine with oxygen atoms in the air and form a new substance, a gas called carbon dioxide. Another example of a chemical change would be the burning of a piece of wood. When wood burns, heat and light are given off. More than just the appearance of the wood has changed during this chemical change; the substances that wood is made of have changed into new substances—carbon dioxide, water, and ash.

Indicators that a chemical change has taken place are a color change, a precipitate forms, fizzing or bubbling occur, a different odor is produced, or heat or light are given off. A precipitate is a new solid that forms when two liquids are mixed together. The Glubber, a precipitate that is produced in this reaction, is a polymer similar to Slime™ and Silly Putty™.

Management
1. Borax is available commercially in grocery stores. It is sold under the name brand 20 Mule Team Borax. Make sure the students treat the Borax solution with care. Borax is toxic if ingested in large quantities. Follow the cautionary statements on the side of the package.

2. Students can store their Glubber mixtures in small zipper-type plastic bags.

3. A saturated borax solution can be prepared by filling a container with one liter of water and adding powdered borax while stirring until no more will dissolve in the water. Demonstrate the technique for making the solution so students will be able to make more if they need it during their investigations.

4. You can make inexpensive containers by cutting the tops off two-liter soda bottles.

5. Demonstrate how to use the graduated cylinder to mark the cup so that glue will not be poured into the graduated cylinder. Begin by filling the graduated cylinder to 10 mL with water and pouring the water into the plastic cup. Mark the water level on the side of the cup with the permanent marker. Pour the water out and measure 20 mL and repeat. Tell the students to create lines for 30 and 40 mL utilizing the same procedure. The cup will now be used as a graduated measuring cup. Point out that one meaning of the word *graduate* means to separate equally.

6. Prepare the task cards. You may want to make a set of double stations if this will enable a smoother rotation through the centers.

7. Each group will need the equivalent of a 4-oz bottle of white glue.

8. Tell the students to keep the "glue" spoon and "borax" spoon separate.

Procedure

Part One

1. Ask the *Key Question* and state the *Learning Goals*.

2. Distribute to each group the two plastic cups, permanent markers, and graduated cylinders.

3. Demonstrate how to use the graduated cylinder to mark the sides of the cup.

4. Allow the students time to mark both their cups.

5. Make a saturated solution of the borax in the container. Point out that particles will settle to the bottom when the water can hold no more borax.

6. Discuss how a percentage describes the formula for making each type of Glubber. The first formula is 50 percent glue and 50 percent water—20 mL of glue and 20 mL of water.

7. Tell each group to mix the glue and water based on the first formula. Pour the glue in first until it reaches the 20-mL mark. Fill with water up to the 40-mL mark. Emphasize the importance of mixing the two ingredients thoroughly with the plastic spoon.

8. Have each group pour 40 mL of the borax solution into the second cup. Direct them to use the second spoon and place an additional

spoonful of borax into the borax solution and stir. (The suspended particles aid in the chemical reaction.)

9. Tell the students to make as many observations about the properties of the two solutions as they are able.

10. Demonstrate how to pour the glue/water mixture into the borax solution while stirring the mixture with the spoon.

11. A glob of Glubber will form on the spoon. Have the students remove the Glubber from the spoon and work it with their hands until all excess water is removed from the Glubber. Direct them to rinse out their cups with water BEFORE they begin exploring the Glubber.

12. Allow a few minutes of free exploration with the Glubber.

13. Tell the students to write a few observations about this new substance. Have them compare it with the two substances from which it came.

14. Tell the students to place the Glubber they have made into one of the plastic bags. Instruct them to mark the bag with the formula.

15. Direct them to make the other two formulas for Glubber. Remind them to clean their cups immediately after making each formula.

Part Two

1. Tell the students they will be conducting a product test on the three different formulas for Glubber. The results of the testing will determine which formula is best for the new product.

2. Distribute the product-testing sheet and inform students that they will record their observations on how each of the formulas responded to the various tests. Urge them to keep accurate records.

3. State the procedure for the rotation through the stations.

4. Have the students submit a written report based on their findings. Tell them they will need to make a recommendation on which of the three formulas they would recommend using for this new product.

5. Direct them to design an ad to sell the product. The ad should emphasize the unique properties of this new product.

Connecting Learning

1. What evidence do you have that a chemical reaction has taken place when you mixed the glue solution with the borax solution? [A precipitate formed.]

2. Which formula(s) did not work very well. [Those that contained less than 40 percent glue did not work very well.]

3. How did your group decide on the best formula?

4. Did every group select the same formula? Why or why not?
5. Why is it important to base product decisions on experimental data?
6. Did personal bias influence your group's decision on what formula you selected? Explain. (It is possible that one test may have influenced a group more than another. This would be a good opportunity to discuss how bias does exist in science.)
7. The Glubber you made is a polymeric solid. It is made up of long chains of molecules wrapped around each other. What observations did you make that would lead you to infer that molecular chains were forming? [As we poured the solutions together and the Glubber began to form in the liquid, we could see that it was "stringy."]
8. What other formulas would you like to try?

Evidence of Learning
1. Listen for student talk as they are discussing the *Connecting Learning* questions.
2. Look for recommendations supported by test data gathered at the stations.
3. Look for listing of properties that describes this new product in the ad.

* Reprinted with permission from *Principles and Standards for School Mathematics*, 2000 by the National Council of Teachers of Mathematics. All rights reserved.

Product Testing

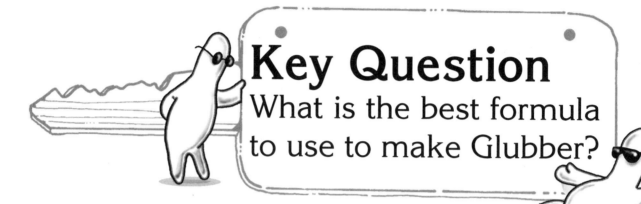

Key Question
What is the best formula to use to make Glubber?

Learning Goals
The students will:
1. identify and utilize percentages,
2. compare and contrast properties of matter, and
3. recommend a product formula based on experimental data.

Cut task cards apart and place at centers.

FORMULA A
20 mL glue
20 mL water
40 mL of borax solution

FORMULA C
40 mL glue
0 mL water
40 mL of borax solution

FORMULA B
30 mL glue
10 mL water
40 mL of borax solution

1

Bounce Test

Drop each Glubber from a height of 100 cm. How high does the Glubber bounce? Sequence the Glubber formulas from highest bounce (1) to lowest bounce (3).

100

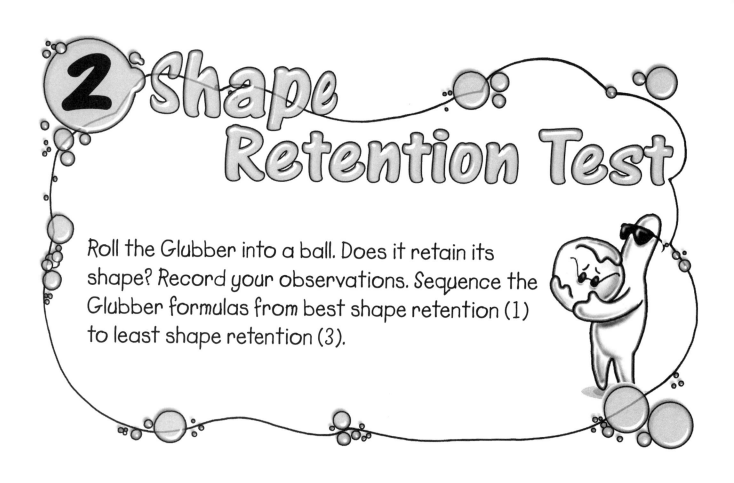

2 shape Retention Test

Roll the Glubber into a ball. Does it retain its shape? Record your observations. Sequence the Glubber formulas from best shape retention (1) to least shape retention (3).

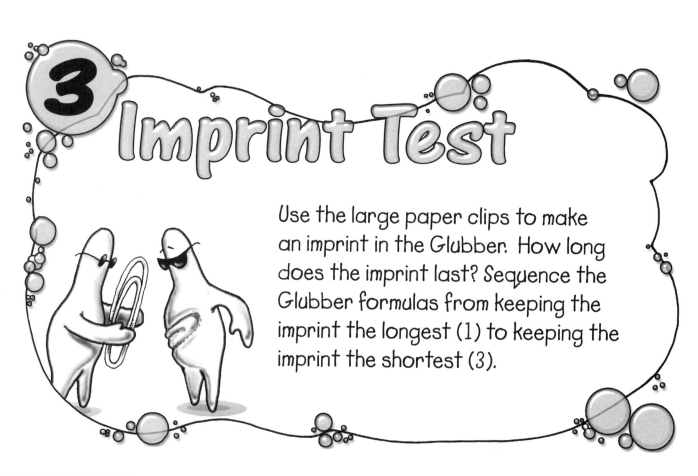

3 Imprint Test

Use the large paper clips to make an imprint in the Glubber. How long does the imprint last? Sequence the Glubber formulas from keeping the imprint the longest (1) to keeping the imprint the shortest (3).

4 Stretch Test

Roll the Glubber into a 20-cm rope. Pull the rope until it snaps. How long was each Glubber rope? Sequence the Glubber formulas from the longest rope (1) to the shortest rope (3).

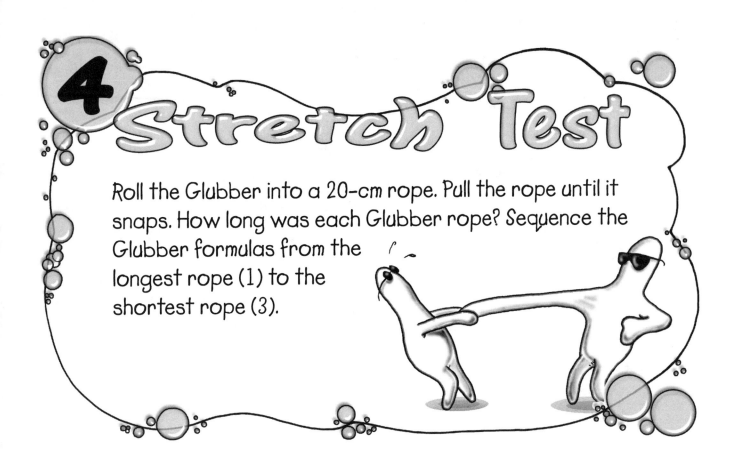

5 Print Transfer Test

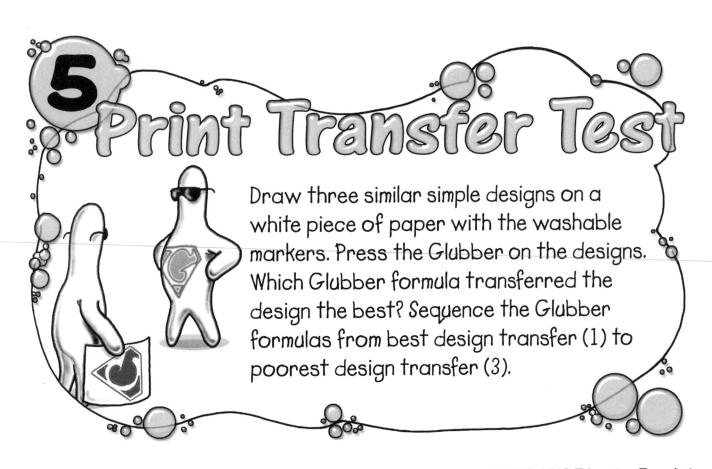

Draw three similar simple designs on a white piece of paper with the washable markers. Press the Glubber on the designs. Which Glubber formula transferred the design the best? Sequence the Glubber formulas from best design transfer (1) to poorest design transfer (3).

Product Testing

Mix each formula with 40 mL of borax solution.

FORMULA A
20 mL glue
_____ %
20 mL water
_____ %

FORMULA B
30 mL glue
_____ %
10 mL water
_____ %

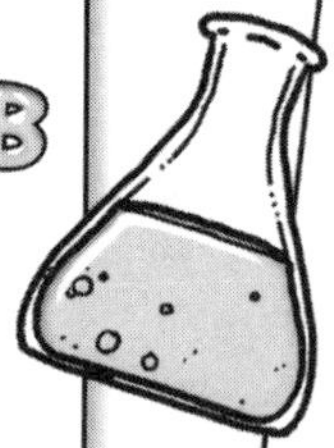

FORMULA C
40 mL glue
_____ %
0 mL water
_____ %

Observation of glue solutions

A.

B.

C.

Observation of borax solutions

A.

B.

C.

Observation of the Glubber

A.

B.

C.

1. How are the solutions the same?

2. How are the solutions different?

3. What happened when you mixed the two solutions?

Product Testing

Record your data from each test.

	Formula A	Formula B	Formula C
Bounce Test			
Shape Retention Test			
Imprint Test			
Stretch Test			
Print Transfer Test			

Product Testing

Give a written summary on which formula you recommend based on the results of the experiments.

Product Testing Recommendation

Formula _____

Design an ad to sell your product. Use the properties of the Glubber as selling features.

Product Testing

Connecting Learning

1. What evidence do you have that a chemical reaction has taken place when you mixed the glue solution with the borax solution?

2. How did your group decide on the best formula?

3. Did every group select the same formula? Why or why not?

4. Why is it important to base product decisions on experimental data?

5. Did personal bias influence your group's decision on what formula you selected? Explain.

6. The Glubber you made is a polymeric solid. It is made up of long chains of molecules wrapped around each other. What evidence did you observe while mixing the two solutions that chains were forming?

7. What other formulas would you like to try?

8. What are you wondering now?

Feel the Heat

Topic
Chemical Reaction

Key Question
What will happen to the temperature of plaster of paris when water is added to it?

Learning Goals
Students will:
1. observe the transformation of chemical energy into heat energy as the temperature increases when water is added to plaster of paris;
2. correctly read a thermometer; and
3. graph and interpret data.

Guiding Documents
Project 2061 Benchmarks
- *Energy cannot be created or destroyed, but only changed from one form into another.*
- *Most of what goes on in the universe—from exploding stars and biological growth to the operation of machines and the motion of people—involves some form of energy being transformed into another. Energy in the form of heat is almost always one of the products of an energy transformation.*
- *Energy appears in different forms. Heat energy is in the disorderly motion of molecules and in radiation; chemical energy is in the arrangement of atoms; mechanical energy is in moving bodies or in elastically distorted shapes; and electrical energy is in the attraction or repulsion between charges.*

NRC Standards
- *In most chemical and nuclear reactions, energy is transferred into or out of a system. Heat, light, mechanical motion, or electricity might all be involved in such transfers.*
- *Heat can be produced in many ways, such as burning, rubbing, or mixing one substance with another. Heat can move from one object to another by conduction.*

NCTM Standards 2000*
- *Represent data using tables and graphs such as line plots, bar graphs, and line graphs*
- *Select and apply appropriate standard units and tools to measure length, area, volume, weight, time, temperature, and the size of angles*

- *Collect data using observations, surveys, and experiments*

Math
Measurement
  time
  temperature
Data analysis
  graphing
    line

Science
Physical science
  matter
    chemistry

Integrated Processes
Observing
Comparing and contrasting
Collecting and recording data
Controlling variables
Inferring
Applying

Materials
Per group:
  3 small paper cups
  craft stick for stirring
  200 mL plaster of paris
  100 mL water
  3 immersible thermometers
  5 cm square of aluminum foil
  clock
  graduated cylinder

Background Information
The *law of conservation of energy* says that energy can neither be created nor destroyed, but it may be converted from one form to another. This activity involves the conversion of chemical energy into heat energy.

An energy change accompanies every change in a substance. Chemical reactions involve one or more substances being changed into one or more new substances.

In this activity, water and powdered plaster of paris are combined into a water-plaster mixture that sets up. The water and plaster of paris contain more chemical potential energy than the water-plaster mixture. Some of the potential energy is lost as the chemical reaction occurs when the two are mixed; this potential energy is converted to heat energy during the reaction. Students will be able to feel the heat being released. Some packages of plaster of paris even carry the warning that the water-plaster mixture should not be used as body casting material because burns may result due to the heat that is generated.

Students should recognize that a chemical reaction has occurred because a change in temperature is one of the indications.

Management
1. Students should work together in groups of four.
2. The bulb end of one thermometer per group should be wrapped in an "envelope" of aluminum foil so that it may be removed without breaking from the hardened plaster of paris.

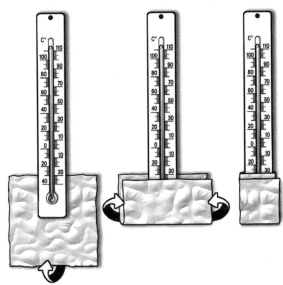

3. The water and plaster of paris should be at room temperature to begin this activity.
4. This activity requires about two hours of periodic observations. You may want to begin it right before students start some other task or leave for a pull-out class. Students should observe the temperatures of the three substances at 10 to 15 minute intervals.

Procedure
1. Discuss different types of energy and transformations of which students are aware. For instance: chemical energy to heat energy which occurs when a uninsulated wire is connected across the terminals of a battery; radiant energy to chemical energy which occurs when the sun's energy is utilized in the photosynthetic process to produce chlorophyll; light energy to heat energy which occurs when a light bulb gives off light

and heat; chemical energy to mechanical energy which occurs when food is converted to muscle movements.
2. Inform students that they are going to keep records of the temperatures of three different substances: water, plaster of paris (powdered form), and the combination of water and plaster of paris.
3. Distribute the three small paper cups, a craft stick, 200 mL plaster of paris, 100 mL water, aluminum foil, and three thermometers to the groups.
4. Have students pour 50 mL of water into one cup, 100 mL of powdered plaster of paris into another cup, and 100 mL of powdered plaster of paris to which they add 50 mL of water in a third cup and stir with a craft stick. Caution them not to use the thermometers as a stirring rod.
5. Direct students to insert a thermometer in the cup of water and one into the cup of powdered plaster of paris. For the third thermometer, have them wrap the bulb end in an "envelope" of aluminum foil. This is to form a little pocket so that the thermometer can be removed from the set plaster of paris without breaking. Direct them to place this thermometer in the water-plaster mixture, no deeper than the top of the foil. Urge them to wiggle the thermometer a little to loosen the foil pocket for easy removal.
6. Make certain they leave the thermometers in each container for about two minutes, or long enough for the thermometer to stabilize, before recording beginning temperatures.

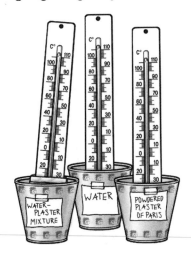

7. After 10 or 15 minutes, have students read and record the temperatures from the three thermometers.
8. Direct them to continue this procedure every 10 or 15 minutes.
9. Encourage students to remove their thermometers and clean them after the data have been collected.
10. Have students construct a graph of their data with *Time* on the x-axis and *Temperature* on the y-axis. Urge them to use three different-colored lines to

represent the temperatures of water, powdered plaster of paris, and the water-plaster mixture. Have them make a key. Direct them to write a conclusion about the energy transformation that occurred.

Connecting Learning

1. What was the temperature of the water before it was mixed? What was the temperature of the plaster of paris before it was mixed? Why do you think they were the same? [They were at room temperature.]

2. When did you begin to notice a temperature change? How is that change reflected on your graph?

3. What was the hottest temperature you recorded? What was the hottest temperature recorded in the class?

4. What was the temperature difference between the dry plaster of paris and the mixture? Was that the same for the entire class? How can you explain any differences?

5. What do you suppose will happen to the temperature of the mixed plaster of paris if we were to leave the thermometers in overnight?

6. What other tests would you like to do with the plaster of paris mixture? (perhaps try larger quantities, add more water or more plaster of paris, add vinegar to the plaster of paris rather than water, etc.)

7. In what real-world situations do you suppose this same thing happens? [in concrete that is drying] Who could we ask to find out if this is true?

8. What other chemical reactions would you like to test to see if this same thing happens? Would it be safe to test them in the classroom?

Evidence of Learning

1. Listen for student explanations during the *Connecting Learning* questions that reference observations made during the investigation

2. Look for accuracy in reading a thermometer during the investigation.

3. Check graphs for correct labeling of x and y-axis as well as accurate plotting of data gathered during the investigation.

Extension

Have students do research about how the heat is removed from huge concrete structures as they cure (dry and harden). It is estimated that it would have taken decades for the temperature of the concrete in the Grand Coulee Dam (on the Columbia River in Washington) to equalize with the surrounding temperatures. To help speed the cooling process, thousands of miles of water pipe where placed in the structure as it was being poured. Cold water was then pumped through the pipes to help remove some of the heat.

* Reprinted with permission from *Principles and Standards for School Mathematics*, 2000 by the National Council of Teachers of Mathematics. All rights reserved.

Key Question

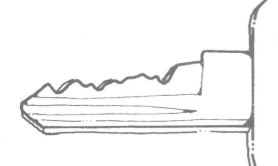

What will happen to the temperature of plaster of paris when water is added to it?

Learning Goals

Students will:

1. observe the transformation of chemical energy into heat energy as the temperature increases when water is added to plaster of paris,
2. correctly read a thermometer, and
3. graph and interpret data.

Feel the Heat

Elapsed Time and Temperature

Substance	min °	min °	min °	min °	min °	min °	min °	min °	min °	min °	min °
Water											
Plaster of paris											
Water and plaster of paris											

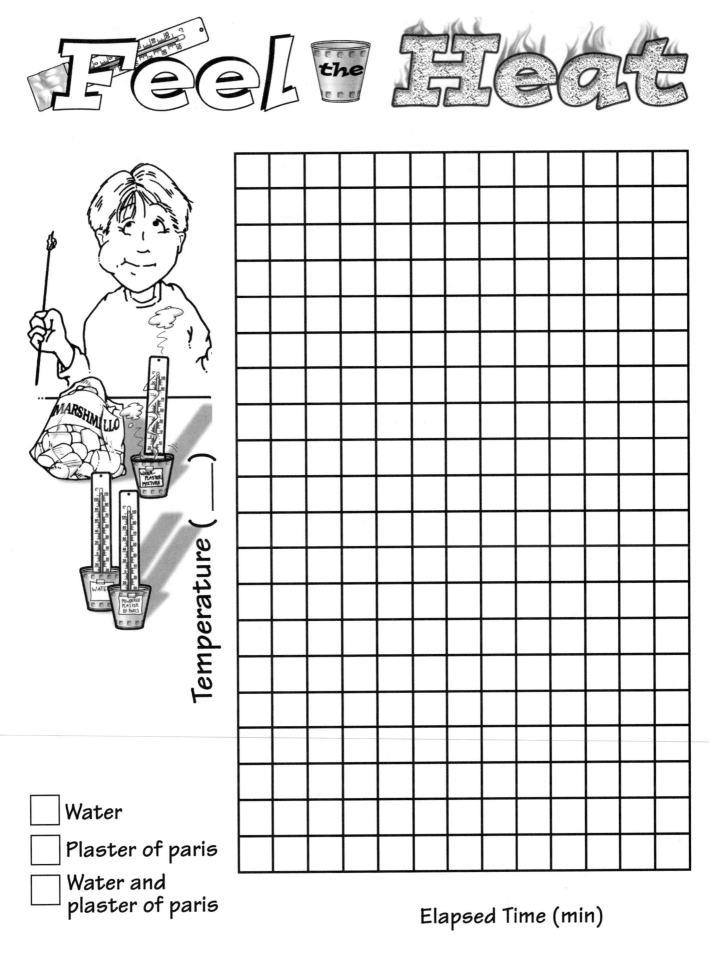

Temperature (___)

☐ Water

☐ Plaster of paris

☐ Water and
plaster of paris

Elapsed Time (min)

Connecting Learning

1. What was the temperature of the water before it was mixed? What was the temperature of the plaster of paris before it was mixed? Why do you think they were the same?

2. When did you begin to notice a temperature change? How is that change reflected on your graph?

3. What was the hottest temperature you recorded? What was the hottest temperature recorded in the class?

4. What was the temperature difference between the dry plaster of paris and the mixture? Was that the same for the entire class? How can you explain any differences?

5. What do you suppose will happen to the temperature of the mixed plaster of paris if we were to leave the thermometers in overnight?

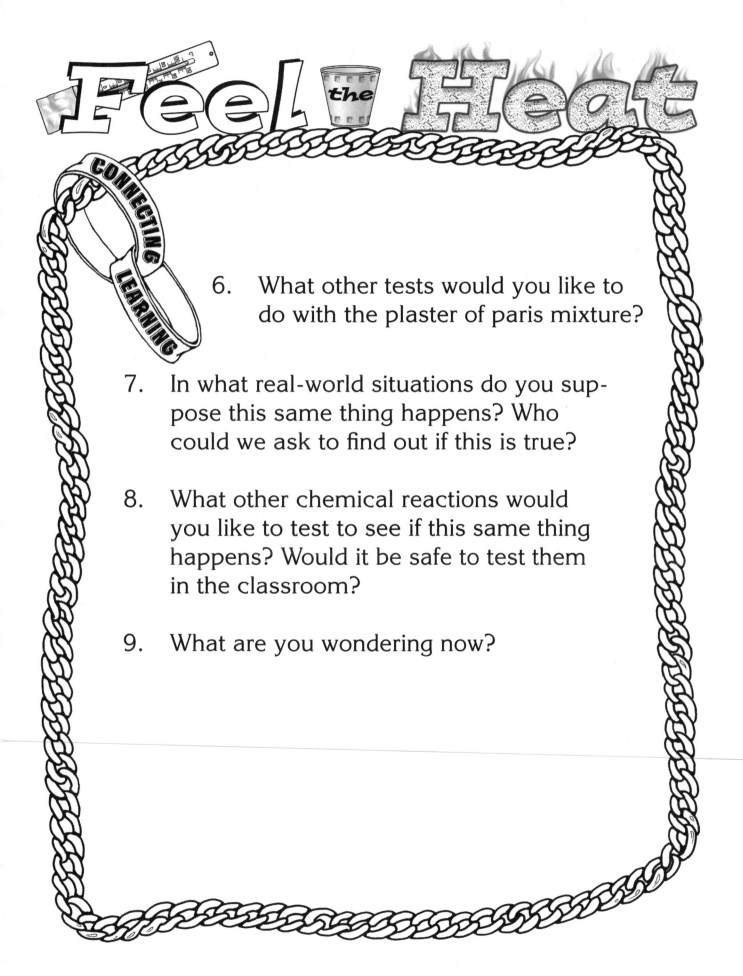

Feel the Heat

CONNECTING LEARNING

6. What other tests would you like to do with the plaster of paris mixture?

7. In what real-world situations do you suppose this same thing happens? Who could we ask to find out if this is true?

8. What other chemical reactions would you like to test to see if this same thing happens? Would it be safe to test them in the classroom?

9. What are you wondering now?

Flipping over Ice Cream

Topic
Matter: Changes, Identifying Properties, Mixtures

Key Question
How can melting ice make our ice cream freeze?

Learning Goals
Students will:
1. freeze individual portions of ice cream, and
2. observe that salt alters the freezing point of water, and
3. infer that the salt and ice mixture will actually cause their ice cream mixture to freeze.

Guiding Documents
Project 2061 Benchmarks
- *Measuring instruments can be used to gather accurate information for making scientific comparisons of objects and events and for designing and constructing things that will work properly.*
- *Heating and cooling cause changes in the properties of materials. Many kinds of changes occur faster under hotter conditions.*
- *When a new material is made by combining two or more materials, it has properties that are different from the original materials. For that reason, a lot of different materials can be made from a small number of basic kinds of materials.*
- *When people care about what is being counted or measured, it is important for them to say what the units are (three degrees Fahrenheit is different from three centimeters, three miles from three miles per hour).*
- *Measure and mix dry and liquid materials (in the kitchen, garage, or laboratory) in prescribed amounts, exercising reasonable safety.*

NRC Standards
- *Objects have many observable properties, including size, weight, shape, color, temperature, and the ability to react with other substances. Those properties can be measured using tools, such as rulers, balances, and thermometers.*
- *Materials can exist in different states—solid, liquid, and gas. Some common materials, such as water, can be changed from one state to another by heating or cooling.*

NCTM Standards 2000*
- *Select and apply appropriate standard units and tools to measure length, area, volume, weight, time, temperature, and the size of angles*
- *Collect data using observations, surveys, and experiments*
- *Represent data using tables and graphs such as line plots, bar graphs, and line graphs*
- *Purpose and justify conclusions and predictions that are based on data and design studies to further investigate the conclusions or predictions*

Math
Measuring
   temperatures
   ingredients
   time

Science
Physical science
   chemistry
    matter

Integrated Processes
Observing
Comparing and contrasting
Collecting and recording data
Interpreting data
Drawing conclusions

Materials
Freezer plastic bags, gallon-size, zipper-type
Freezer plastic bags, pint-size, zipper-type
Ice, crushed or small cubes
Salt
3 thermometers
Ice cream ingredients and equipment,
   see *Recipe Card*
Large bowl
Measuring cups, 1 cup and 1/2 cup
Measuring cup with handle, 1/3 cup
Mixer, electric or rotary hand beater
2 clear plastic cups, 9-oz.
Plastic spoons
Newspaper
Paper towels
Permanent markers
Mittens or gloves, optional

Background Information

Often a great surprise to students is to find that freezing points vary with different substances. They readily associate freezing as being 0°C (or 32°F). Weather forecasters report freezing temperatures as 0°C or 32°F because that is the temperature at which precipitation (water) freezes. Because of this common, real-world occurrence, students assume that this temperature is THE freezing point. This activity will help to dispel this notion.

When students go to freeze their ice cream mixture, they will put it into a ice/salt mixture which has a lower freezing point than ice (-18°C or 0°F). Because the salt and ice get colder than the freezing point of water (the main ingredient in the milk that is in ice cream), the ice cream will freeze. This will be quite discrepant to students because they see the ice in the ice/salt mixture melting and think that it is getting warmer instead of colder.

Management

1. This activity has students compare
   * the temperature of the ice cream mix before freezing with the temperature of the mix after freezing, and
   * the temperature of the ice with the temperature of the ice/salt mixture.

   The teacher or a student should make an extra "sample" bag of ice cream in order to find the temperature of the mixture before and after freezing. Discussions should occur about the changes in the states of matter for the ice cream mixture and for the ice and the ice/salt mixture.
2. Each group of four students will need a gallon-size plastic bag. Each individual student will need a pint-size plastic bag and a plastic spoon.
3. It is important that you use freezer-quality plastic bags to help prevent breaking while the students are flipping the bags to freeze the ice cream.
4. The pint-sized bags should be labeled with the students' names or numbered one through four for each member of the group. This way students will be assured of getting their bag of ice cream after it is frozen.
5. To capitalize on the opportunity for real-world measurement, the entire class can have a hand in preparing the ice cream recipe. If time does not permit this, the mixture can be made beforehand and stored in the refrigerator until time to be used. It is best to keep the ingredients cold in order to speed the freezing process. One recipe is enough for 25 – 30 students if using slightly less than 1/3 cup mixture per student.
6. When students are ready to freeze the ice cream, have them put several layers of newspaper on the table surface to absorb the condensation that forms on the outside of the bag.

7. Mittens for the students are optional. If they grab just the corners of the bag to flip it, their hands will not get uncomfortably cold. Since there are four students in a group, each member can take a stint at flipping the bag.
8. Be aware of any students that may not be able to tolerate foods with sugar and/or milk. Have an alternative snack for these students when others are enjoying the ice cream they have made.
9. Check with your custodian to find out how to dispose of the saltwater mixture.
10. Although Celsius is the preferred temperature scale for science-related activities, Fahrenheit can be used if that is the calibration used on your thermometers. You will need to scale the graphs before copying the student page or work through the scale with the class as they record their data.

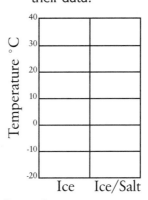 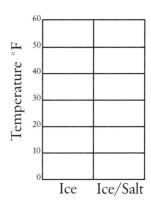

Procedure

1. Ask the students if they have ever made their own ice cream. Discuss what it takes to make ice cream—the various ingredients and cold temperatures.
2. Question them as to whether they think they could make ice cream without using the freezer portion of a refrigerator. Record their ideas for doing this.
3. Bring the discussion to the place where students understand that ice sitting out at room temperature warms up and melts. Probe the students on how they think they could use ice to freeze ice cream.
4. Inform them that they will need to have temperatures at least as low as 0°C (or 32°F) in order to freeze the ice cream and that a part of this activity's purpose is to discover a way to make the ice colder than 0°C (or 32°F). Tell them that they will add something to the ice that will cause that change. Ask for ideas as to what that addition may be.
5. If no one comes up with the idea of adding salt, inform them that this is what will be used.
6. Invite a student to set up two cups to use as temperature data for the whole class. Direct the student to put a thermometer into a cup of

ice and another thermometer into a cup of ice that has a tablespoon of salt sprinkled on it and gently stirred.

7. Allow these cups and thermometers to sit while the ice cream is made. Meanwhile go through the procedure found on the recipe card for making and freezing the ice cream.

8. Divide the class into groups of four. Have students make the ice cream mixture or use already prepared mixture and let students begin freezing. Use a thermometer to determine the temperature of the ice cream mixture in the "sample" bag. Record this as "before freezing temperature" on the chalkboard so that students can add it to their activity sheet when they are finished with their ice cream.

9. When the freezing time is finished (approximately 10 – 15 minutes), check the "sample" bag to see if it is frozen. To do this, carefully remove it from the ice/salt, use a paper towel to thoroughly wipe the liquid from the zippered end of the bag (prevents salt water from contaminating the ice cream), and open the bag to see if the mixture is frozen. If it isn't, direct students to continue flipping for a few more minutes. If it is frozen, continue with the procedure.

10. Draw the students attention to the "sample" bag. Inform them that they need to determine the temperature of the frozen mixture. Have a student insert a thermometer and wait two minutes for it to stabilize. Have a student read the temperature and record it on the chalkboard as "after freezing temperature."

11. Now let everyone enjoy their frozen treat.

12. Have students check the temperatures of the ice and the salt/ice mixture. Direct them to fill in the data on their activity sheet and complete the graphs.

13. Follow with a discussion about states of matter, freezing temperatures, etc. Have students finish the questions at the bottom of their activity sheet.

Connecting Learning

1. What did you like best about this activity?
2. What things surprised you?
3. How did the temperature of the ice change once you put salt on it? [The temperature went down.]
4. Explain in your own words why you were able to freeze the ice cream with salt and ice.
5. What matter in this activity changed state? [The ice cream mixture went from liquid to solid. The ice went from solid to liquid.]
6. Why do weather forecasters call 0°C or 32°F freezing? [That's the freezing temperature of water.]

7. If your parents were to ask you what you learned from this experience, what would you tell them?
8. What would you do differently if you were to do this again?
9. What are you wondering now?

Extensions

1. Have students design an investigation to determine whether more or less salt added to ice will change its temperature.
2. Have students try different recipes for the ice cream. Instead of vanilla pudding, have them try chocolate. They could also add real fruit to the ice cream mixture. Whatever they try, encourage them to write a recipe. Students could taste test several recipes and rate them using a scale of one to five with five being delicious!
3. Use the experience to highlight the process of condensation. Water from the atmosphere condenses on the outside of the cold freezer bag.
4. Put the two cups—one with the melted ice and the other with melted ice and salt—into a freezer. Time how long it takes for them to freeze.

* Reprinted with permission from *Principles and Standards for School Mathematics,* 2000 by the National Council of Teachers of Mathematics. All rights reserved.

Flipping OVER ICE CREAM

Key Question

How can melting ice make our ice cream freeze?

Learning Goals

Students will:

1. freeze individual portions of ice cream,
2. observe that salt alters the freezing point of water, and
3. infer that the salt and ice mixture will actually cause their ice cream mixture to freeze.

Flipping OVER ICE CREAM

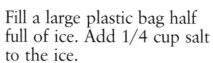

Ice Cream Recipe

2 1/2 cups sugar
1 can (12 oz) evaporated milk
2 teaspoons vanilla
1 package (3.4 oz) vanilla instant pudding mix
6 cups milk

Combine all the ingredients in a large bowl and mix until the sugar and pudding are dissolved.

Place a little less than 1/3 cup of ice cream mixture into a small plastic bag. Zip the bag. Make sure it is sealed tightly.

Fill a large plastic bag half full of ice. Add 1/4 cup salt to the ice.

Put four small, sealed bags of ice cream mixture into the bag with the ice and salt. Seal the large bag.

Place the bag on a layer of newspapers. Grab two corners of the bag and flip it.

Each person in your group should flip the bag end over end for two to three minutes until the ice cream is frozen.

When the ice cream is frozen, carefully remove the small bags from the large bag of ice. Use a paper towel to wipe off your small bag.

Open the bag. Use a spoon to eat the ice cream right out of the bag.

Flipping OVER ICE CREAM

Record the following temperatures:

Ice _____

Ice and salt _____

Ice cream mix (before freezing) _____

Ice cream mix (after freezing) _____

What do the graphs tell you?

Record your data on the graphs.

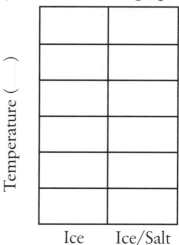

Temperature (__)

| Ice | Ice/Salt |

Ice Cream Mixture

Temperature (__)

| Before Freezing | After Freezing |

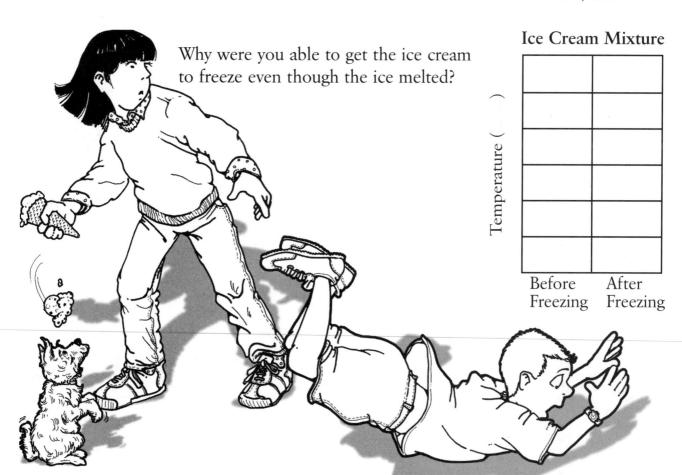

Why were you able to get the ice cream to freeze even though the ice melted?

On the back of this paper, explain what is wrong with the following statement:
The freezing temperature is 0°C or 32°F.

Connecting Learning

1. What did you like best about this activity?

2. What things surprised you?

3. How did the temperature of the ice change once you put salt on it?

4. Explain in your own words why you were able to freeze the ice cream with salt and ice.

5. What matter in this activity changed state?

6. Why do weather forecasters call 0°C or 32°F freezing?

7. If your parents were to ask you what you learned from this experience, what would you tell them?

8. What would you do differently if you were to do this again?

9. What are you wondering now?

Change Matters

Topic
Physical and Chemical Changes

Key Question
How do you know whether the changes you observe are physical or chemical changes?

Learning Goals
Students will:
1. identify physical and chemical changes;
2. classify changes in matter; and
3. collect, graph, and analyze data.

Guiding Documents
Project 2061 Benchmarks
- *Heating and cooling cause changes in the properties of materials. Many kinds of changes occur faster under hotter conditions.*
- *Offer reasons for their findings and consider reasons suggested by others.*
- *Understand writing that incorporates circle charts, bar and line graphs, two-way data tables, diagrams, and symbols.*

NRC Standard
- *Scientists develop explanations using observations (evidence) and what they already know about the world (scientific knowledge). Good explanations are based on evidence from investigations.*

NCTM Standards 2000*
- *Collect data using observations, surveys, and experiments*
- *Represent data using tables and graphs such as line plots, bar graphs, and line graphs*
- *Compare different representations of the same data and evaluate how well each representation shows important aspects of the data*

Math
Data analysis
  Venn diagram

Science
Physical science
  matter

Integrated Processes
Observing
Comparing and contrasting
Collecting and recording data
Inferring
Applying

Materials
Ice cubes
5 transparent plastic cups
Steel wool
Tongs
Vinegar
Paper towels
Plastic spoon
Empty 35 mm film canister
Baking soda
Vinegar
Serrated plastic knife
Apple
Match
Candle
Clay
Aluminum foil
Container of water
Index card for each student

Background Information
Matter can go through physical and chemical changes. In a physical change, the chemical properties of a substance remain unchanged. Melting ice and tearing paper or cloth are examples of physical changes. In each example, the same substance remains—water, paper, and cloth. The form of the substances may have changed, but not the substances' chemical compositions.

In a chemical change, the chemical properties of a substance are changed. Burning a match, photosynthesis, and rusting iron are all chemical changes. In each case the chemical composition has been altered. In a chemical change, the bonds holding the molecules are broken and the atoms reform into different molecules.

Management
1. CAUTION: Make certain that safety precautions are strictly followed at the station where an open flame is used.

118

2. Divide students into small groups to rotate through the stations. To save time, you may want to set up more than one set of stations.

3. The materials required are as follows:
   Station 1: Ice cube, plastic cup
   Station 2: Steel wool pieces, plastic cup with vinegar, tongs, paper towel
   Station 3: Plastic spoon, plastic cup, empty film canister, baking soda, vinegar
   Station 4: Apple, plastic knife, plastic cup
   Station 5: Match, candle, clay, aluminum foil, container of water

4. The analysis of the observations are:
   Station 1: Melting ice cube (This is a physical change, the water is only changing state.)
   Station 2: Steel wool in vinegar (This is an oxidation reaction. The iron in the steel wool bonds with oxygen to form iron oxide (rust). This is a chemical change.)
   Station 3: Baking soda and vinegar (This is a chemical reaction and shows a chemical change. The production of bubbles is one indication of a chemical reaction.)
   Station 4: Sliced apple (Slicing the apple is a physical change. The apple turns brown as oxygen reacts with it which is a chemical change (oxidation).)
   Station 5: Burning candle (The melted wax is a physical change. The burning wick and wax is a chemical change.)

5. It is assumed that students have had some prior experience with physical and chemical changes.

6. Have the index cards ready for the end of the activity. These cards are called an Exit Slip. Use these brief assessments to check for students reasoning for selecting a data organizational tool.

Procedures

1. To review and reinforce, ask students to determine whether the following scenarios are examples of physical or chemical changes and to explain why.
   - They cut their hair. (Physical—nothing new is formed)
   - They burn their toast. (Chemical—the black "stuff" on the toast is a new substance)
   - They dissolve a seltzer tablet in water. (Chemical—bubbles are produced because a gas was formed)
   - They break some rocks with a hammer. (Physical—they are still rocks, just different sizes and shapes)
   - Their ice cream melts. (Physical—still ice cream, but in a different state)
   - They burn a match. (Chemical—the black "stuff" is a new substance)

2. Ask the *Key Question* and state the *Learning Goals*.

3. Distribute recording sheet. Explain the rotation strategy through the stations. Urge students to read the *Station Card* at each station before performing the activity.

4. Direct them to record whether the change they observe is *Physical, Chemical,* or *Both* and to explain why.

5. When students have completed all stations, ask them to return to their seats to discuss their results.

6. Distribute the Venn diagram sheet. Direct students to record their observations.

7. Ask them which format, chart or Venn diagram, they best like to use and why.

8. Ask which format they think provides the most information and why.

9. Ask students to write two examples of physical changes and two examples of chemical changes on their Venn diagram sheet. Have them place descriptors of the changes in the appropriate areas of the Venn diagram.

10. Have students exchange papers to analyze the additional data. Ask if they agree or disagree with the new data that were added to the Venn.

11. Distribute the index cards and have each student write for five minutes on what format, chart or Venn diagram, they liked the best. Tell them they must give reasons for their choice.

Connecting Learning

1. What are the characteristics of a physical change? [the substance can change state, size, or shape, but it is still the same "stuff"]

2. What are the characteristics of a chemical change? [new substances are formed; evidence that a chemical change has occurred may be the production of heat and light, a change in color, the giving off of a gas]

3. If you were to crack an egg and stir it, would this be a physical or chemical change? Explain. Where would it go in the Venn diagram?

4. What if you were to cook that egg? Would that be a physical or chemical change? Explain. Where would it go in the Venn diagram?

5. Did everyone come up with the same results? If not, describe the problems you had in determining whether things at the stations were physical or chemical changes.

6. What are you wondering now?

Evidence of Learning

1. Listen for student discussion during the connecting learning questions. Listen for understanding based on knowledge gained from the activity.

2. Look for accuracy in completing the student sheets. Check student written work for explanations based on observations from the activity.

3. Read each student's Exit Slip to check for the reasoning for selecting the data tool they liked the best.

* Reprinted with permission from *Principles and Standards for school Mathematics*, 2000 by the National Council of Teachers of Mathematics. All rights reserved.

Change Matters

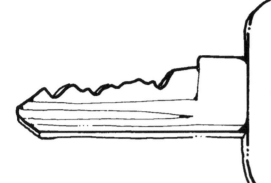

Key Question

How do you know whether the changes you observe are physical or chemical changes?

Learning Goals

STUDENTS WILL:

1. identify physical and chemical changes;
2. classify changes in matter; and
3. collect, graph, and analyze data.

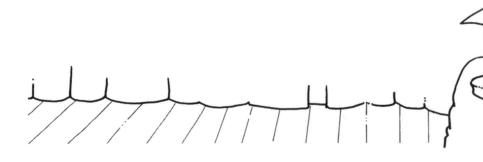

Change Matters

Record your observations.

Station	Physical	Chemical	Both	Why?
1. Ice cube				
2. Steel wool in vinegar				
3. Baking soda and vinegar				
4. Apple slice				
5. Burning candle				

Change Matters

In the Venn diagram, indicate the proper placement of each change.

Chemical

Physical

CHANGING ROOM ☆

1. Ice cube
2. Steel wool in vinegar
3. Baking soda and vinegar
4. Apple slice
5. Burning candle

Change Matters

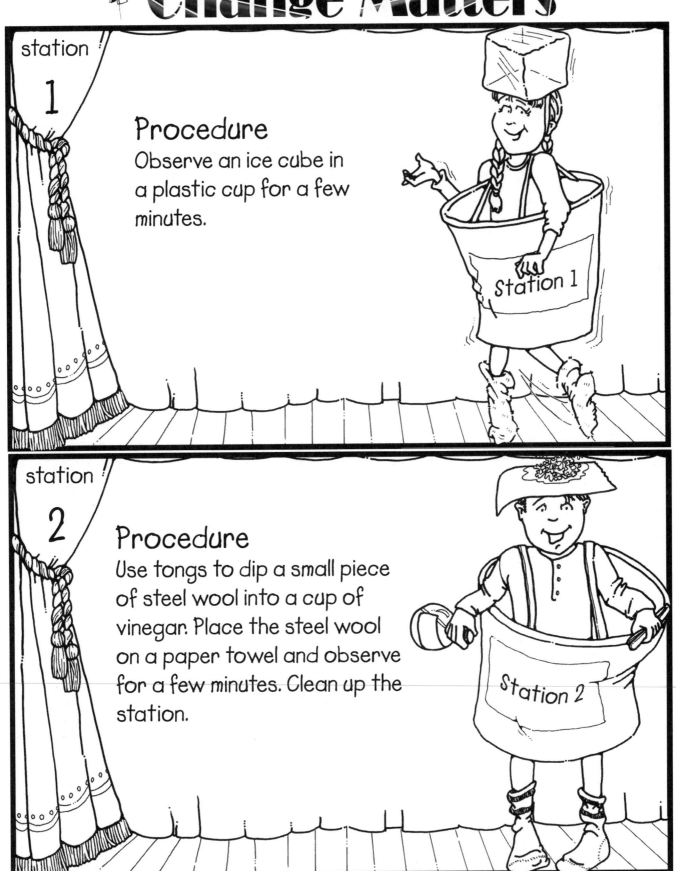

station 1

Procedure
Observe an ice cube in a plastic cup for a few minutes.

Station 1

station 2

Procedure
Use tongs to dip a small piece of steel wool into a cup of vinegar. Place the steel wool on a paper towel and observe for a few minutes. Clean up the station.

Station 2

station 3

Procedure

Place a spoonful of baking soda into the plastic cup. Fill the film canister with vinegar and pour it into the cup with baking soda. Observe. Clean up the materials before leaving the station.

station 4

Procedure

Cut a small wedge from the apple and place it in the plastic cup. Observe it for a few minutes.

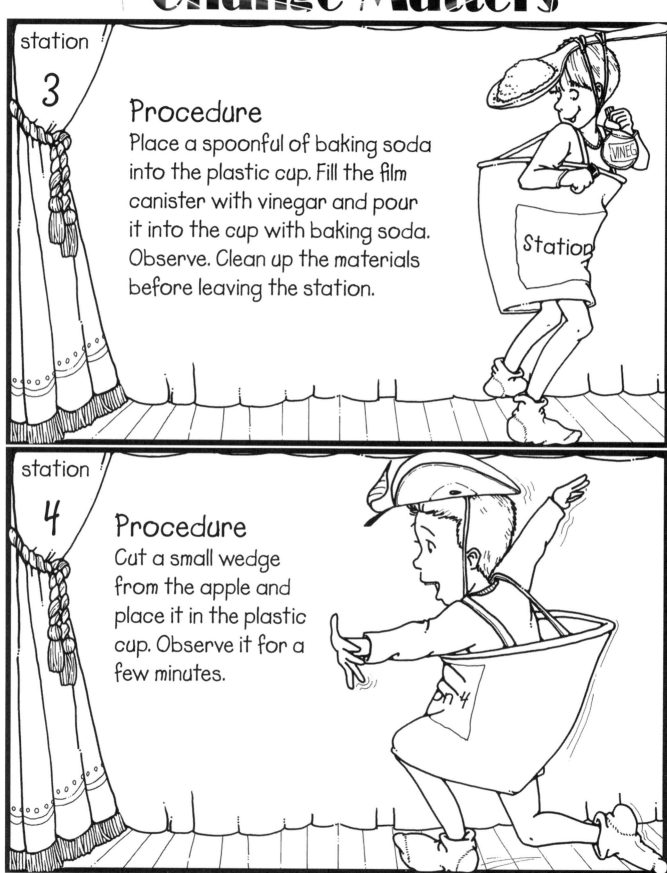

Change Matters

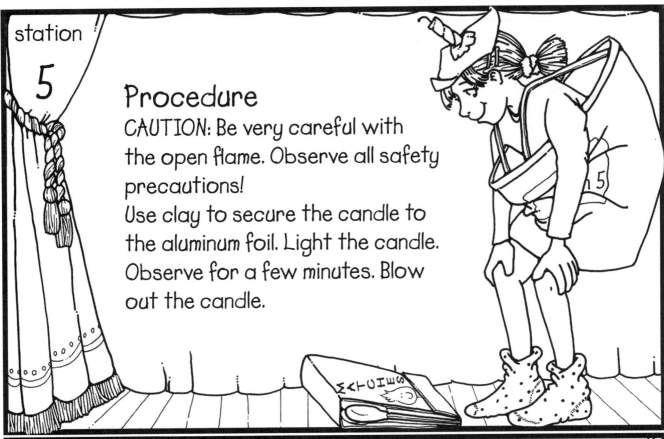

Procedure

CAUTION: Be very careful with the open flame. Observe all safety precautions!

Use clay to secure the candle to the aluminum foil. Light the candle. Observe for a few minutes. Blow out the candle.

Materials needed for each station:

Station 1: Ice cube, plastic cup
Station 2: Steel wool, plastic cup with vinegar, tongs, paper towel
Station 3: Plastic spoon, plastic cup, empty film canister, baking soda, vinegar
Station 4: Apple, plastic knife, plastic cup
Station 5: Match, candle, clay, aluminum foil, container

Change Matters

Connecting Learning

1. What are the characteristics of a physical change?

2. What are the characteristics of a chemical change?

3. If you were to crack an egg and stir it, would this be a physical or chemical change? Explain. Where would it go in the Venn diagram?

4. What if you were to cook that egg? Would that be a physical or chemical change? Explain. Where would it go in the Venn diagram?

5. Did everyone come up with the same results? If not, describe the problems you had in determining whether things at the stations were physical or chemical changes.

6. What are you wondering now?

SELECTIVE SERVICE

Topic
Osmosis, Selectively-Permeable Membranes

Key Question
How can an egg show us how materials are transported from place to place?

Learning Goals
Students will:
1. observe the evidence of a chemical reaction,
2. observe the passing of materials through a membrane, and
3. relate this process to transport systems in living organisms.

Guiding Documents
Project 2061 Benchmarks
- *One way to make sense of something is to think how it is like something familiar.*
- *Use numerical data in describing and comparing objects and events.*

NRC Standards
- *Scientific investigations involve asking and answering a question and comparing the answer with what scientists already know about the world.*
- *Use appropriate tools and techniques to gather, analyze, and interpret data.*
- *Objects have many observable properties, including size, weight, shape, color, temperature, and the ability to react with other substances. Those properties can be measured using tools, such as rulers, balances, and thermometers.*

NCTM Standards 2000*
- *Understand such attributes as length, area, weight, volume, and size of angle and select the appropriate type of unit for measuring each attribute*
- *Collect data using observations, surveys, and experiments*
- *Represent data using tables and graphs such as line plots, bar graphs, and line graphs*
- *Describe the shape and important features of a set of data and compare related data sets, with an emphasis on how the data are distributed*

Math
Measurement
  circumference
  mass
Estimation
Graphing

Science
Physical science
  chemistry
    chemical reactions
  osmosis
  selectively-permeable membranes
Life science
  transport of materials

Integrated Processes
Observing
Comparing and contrasting
Collecting and recording data
Analyzing data
Drawing conclusions
Applying

Materials
Eggs (see *Management 3*)
Vinegar (see *Management 5*)
Plastic cups (see *Management 4*)
Light corn syrup
Water
String, 30 cm per group
Balances
Metric masses
Meter tapes
Paper towels
Labels (see *Management 6*)

Background Information
    In all living organisms, the materials involved in such processes as respiration, nutrient transfer, waste removal, etc. must be able to move from place to place. In order to do this, they have to be able to move across membranes. It is also necessary that these membranes allow only specific materials to cross them; they need to be selective. A selectively permeable membrane is one that allows only certain materials to cross it. These membranes are the gatekeepers that let things in or out.

    Selectively permeable membranes, also called semi-permeable membranes, make it possible for minute particles to pass through the walls of capillaries, intestine villi, and cell membranes. Some examples in extreme simplicity: Oxygen from the lungs passes through capillary walls to be transported by the blood to all parts of the body. Carbon dioxide passes from the cells through the capillaries and is transported by the blood to the lungs to be exhaled. Broken down food particles in the small intestine pass from the intestinal villi into the blood stream to be transported throughout the body to nourish it.

In this experience, students will use vinegar to decalcify an egg. They should recognize that the bubbles that are produced are evidence (evolution of a gas) that a chemical reaction is occurring between the acid in the vinegar and the calcium in the eggshell. Once the egg has been decalcified, the egg is left intact with just the rubbery membrane holding it together. This membrane is a selectively-permeable membrane that allows only particles of a certain size to pass through.

Some of the decalcified eggs will be placed in corn syrup. After a short period of time, students will notice that the egg is shrinking in mass and circumference. This is because of osmosis—the passing of water molecules through the selectively-permeable membrane from areas of high concentration of water into areas of lower concentration. This higher water content of the egg means that its water will pass through the membrane into the corn syrup. Have students notice that the yolk does not pass through this membrane.

Students will also place some of the decalcified eggs into a container of water. Because the container of water is of a higher water concentration than the water in the egg, the water from the container will pass through the egg's membrane into the egg. Students will notice that the egg gains in both mass and circumference.

Management

1. This activity is divided into three parts. *Part One*—the decalcification of the eggs. *Part Two*—the passing of water from the egg into the corn syrup and the passing of water from the container into the egg. *Part Three*—the playing of a Red Rover-type game with students to reinforce the idea of a selectively-permeable membrane.
2. Divide the class into groups of three of four students.
3. Each group will need at least two decalcified eggs. You may want to have extras available in case of breakage. Warn students that they must be careful with the eggs; even though the membranes are tough and flexible, they do break.
4. Each group will need as many plastic cups as they have eggs.
5. White vinegar from the grocery store works well. You may want to cover the tops of the vinegar-filled cups with plastic wrap in order to keep the odor down.

6. Copy enough labels, *Pass on Through* and *Don't Pass Through,* so that each student can have one. Cut the labels apart for use when doing *Part Three.*
7. The string will be used for measuring the circumference of the eggs. Students will put the string around the "equator" of the egg, mark the spot of overlap, and then compare the length to the centimeter measures on the meter tape. This will prevent getting corn syrup, etc. on the meter tapes.

Procedure
Part One
1. Ask the *Key Question* and state the *Learning Goals.*
2. Distribute at least two eggs and two plastic cups to each group. Tell students to write *Egg 1* and *Egg 2* on a piece of transparent tape or masking tape to use as labels for the cups. Students may also want to put a group name on the tape to insure that they get the same eggs with which they began.
3. Have students carry their cups and eggs to a designated place where they will sit in vinegar for about two days.
4. Once the eggs are in place, let students add vinegar to their cups, telling them to add enough to cover the eggs.
5. After two days, review the *Key Question* and the *Learning Goals.*
6. Distribute the first student page. Go over the directions on the page. Ask the students what things they will have to collect in order to complete the page. [eggs and vinegar, balance, masses, string, meter tape, paper towels] Write these on the chalkboard or on the overhead projector so that students can be sure they have gotten all their supplies. Give them two or three minutes to gather everything.
7. Once everyone is back in their seats, ask students what they observe about the eggs and vinegar. [The shells are gone, there is white stuff and bubbles in the vinegar.] Ask them what evidence there is that a chemical reaction has occurred.
8. If there is any shell left on the egg, tell students to gently rub if off. Direct them to drain the vinegar, wash and dry the cups, and rinse the egg. Have them complete the information on the rest of the page. Describe how they should use the string to measure the egg's circumference.

Part Two
1. Distribute the second student sheet. Have students follow the directions for putting one egg in water and one in corn syrup.
2. Inform them that when they predict what will happen, that they may not be making a prediction because they don't have prior experience on which to build a prediction, but they may just be making a wild guess which is acceptable in this situation.

129

3. Have the groups assign one member to be the timekeeper to make sure they are making measurements every ten minutes. Direct them to record the *Start* time mass and circumference from the actual measures they recorded on the first student sheet.

4. After the entire 30-minute period, have students complete the graphs on the third student page and write their conclusions. Make certain that they color code the data by using the color key.

5. Discuss how the selectively-permeable (permeable means some things pass through) membrane allowed something to pass through from the egg into the corn syrup, and the water to pass through into the egg. The membrane was selective in what it allowed to go through.

6. Discuss with the students how living organisms have selectively-permeable membranes. This is how water is absorbed by the root hairs of plants. It is how nutrients are absorbed into the blood stream. It is how oxygen and carbon dioxide get into the blood stream.

Part Three
1. Tell the students that they are going to play a modified version of Red Rover and apply it to selectively-permeable membranes.

2. Divide the class into two teams. Explain that they will form two lines, facing each other and holding hands.

3. Have one side start by call a person from the opposing team to "(Name, Name) we want you to try to pass through." That person then runs toward the other line trying to break through the opposing team. If he or she breaks through, the student selects the label *Pass on Through* and chooses a member from the opposing team to join his or her team. If he or she fails to break through, the runner selects the label *Don't Pass Through* and joins the other team.

4. Try to continue the game long enough for every student to have a chance to be the runner.

5. Ask the students how this game is like the egg experiment. What is the significance of the labels?

Connecting Learning
1. What evidence did you have that a chemical reaction took place?

2. What was the result of this chemical reaction? [the acid in the vinegar reacted with the calcium and the egg was left without its shell]

3. When you rolled the decalcified egg across your arm and across the paper towel, what did you notice? [It left a wet trail.] Why do you think this happened? [The water in the egg was able to pass through the egg's membrane.]

4. Explain what you observed when you put the decalcified egg in the corn syrup. ...in the water.

5. Why did this happen? [The membrane of the egg lets some things pass through.]

6. What are some examples of selectively-permeable membranes in living things?

7. Explain in your own words how the game of Red Rover is like what happened with the egg.

8. What things passed through the egg's membrane that would be like the students who got the labels *Pass on Through?*

9. What would be something that *Does not Pass Through?*

10. What are you wondering now?

Extensions
1. Let students see what happens if they leave their eggs in corn syrup and water for longer periods of time.

2. Let students select other liquids for their decalcified eggs. What about rubbing alcohol, plaster of Paris, water with food coloring, salt water, sugar water, cola, milk, etc.?

* Reprinted with permission from *Principles and Standards for School Mathematics,* 2000 by the National Council of Teachers of Mathematics. All rights reserved.

SELECTIVE SERVICE

Key Question

How can an egg show us how materials are transported from place to place?

Learning Goals

STUDENTS WILL:

1. observe the evidence of a chemical reaction,

2. observe the passing of materials through a membrane, and

3. relate this process to the transport systems in living organisms.

PART ONE

SELECTIVE SERVICE

1. First, soak raw eggs in vinegar for about two days. What do you notice about the eggs in vinegar?

2. What do you think is happening to the eggs?

3. Take out one decalcified egg. Handle it with care! Gently blot it dry. Roll it along the chalkboard, a paper towel, and your arm. What do you notice?

4. Gently blot two decalcified eggs with a paper towel. Estimate their masses and circumferences. After you've estimated them, find their actual measures.

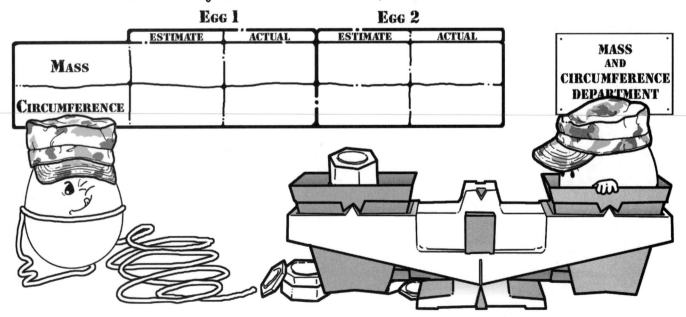

	EGG 1		EGG 2		
	ESTIMATE	ACTUAL	ESTIMATE	ACTUAL	MASS AND CIRCUMFERENCE DEPARTMENT
MASS					
CIRCUMFERENCE					

132

SELECTIVE SERVICE

1. Place one decalcified egg into one cup and the second egg into another cup. Fill the first cup with water and the second cup with corn syrup.
Describe your observations.

2. Predict what you think will happen.

3. Soak the eggs for a total of 30 minutes. Take them out at 10-minute intervals to find their circumferences and masses. Record your information in the tables.

MASS

	START	10 MIN.	20 MIN.	30 MIN.
EGG IN WATER	g	g	g	g
EGG IN SYRUP	g	g	g	g

CIRCUMFERENCE

	START	10 MIN.	20 MIN.	30 MIN.
EGG IN WATER	cm	cm	cm	cm
EGG IN SYRUP	cm	cm	cm	cm

133   © 2003 AIMS Education Foundation

PART TWO

$\mathbf{S}$ELECTIVE $\mathbf{S}$ERVICE

COLOR KEY

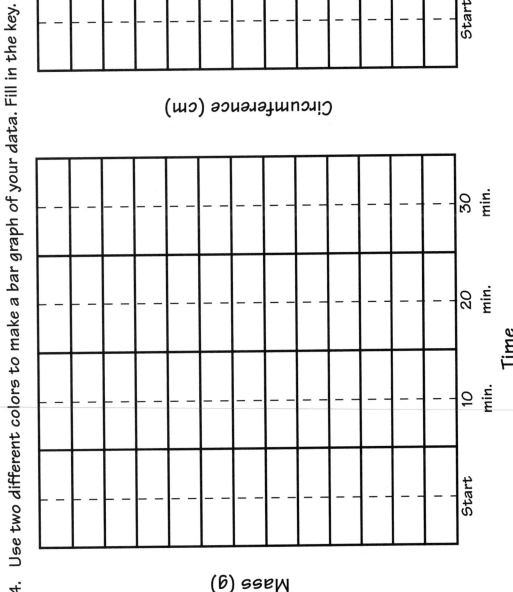

= EGG IN WATER

= EGG IN SYRUP

4. Use two different colors to make a bar graph of your data. Fill in the key.

Circumference (cm)

Start | 10 min. | 20 min. | 30 min.

Time

Mass (g)

Start | 10 min. | 20 min. | 30 min.

Time

5. Carefully analyze your results. What is your conclusion? (Use the back of this paper.)

PART THREE

SELECTIVE SERVICE
RED ROVER

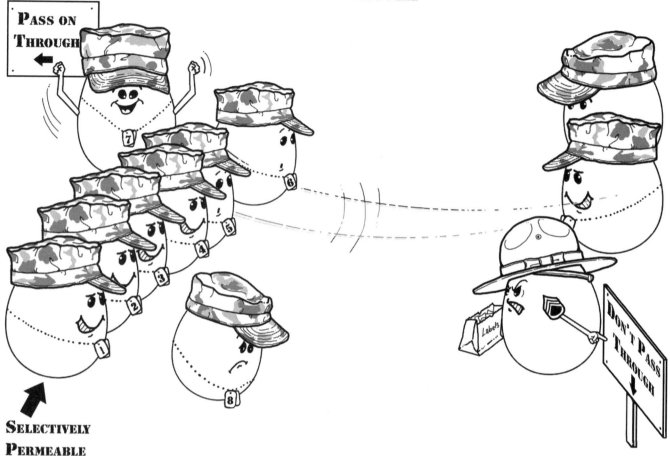

PASS ON THROUGH ←

SELECTIVELY PERMEABLE MEMBRANE

Labels

DON'T PASS THROUGH ↓

How is this game of Red Rover like the eggs in corn syrup and water?

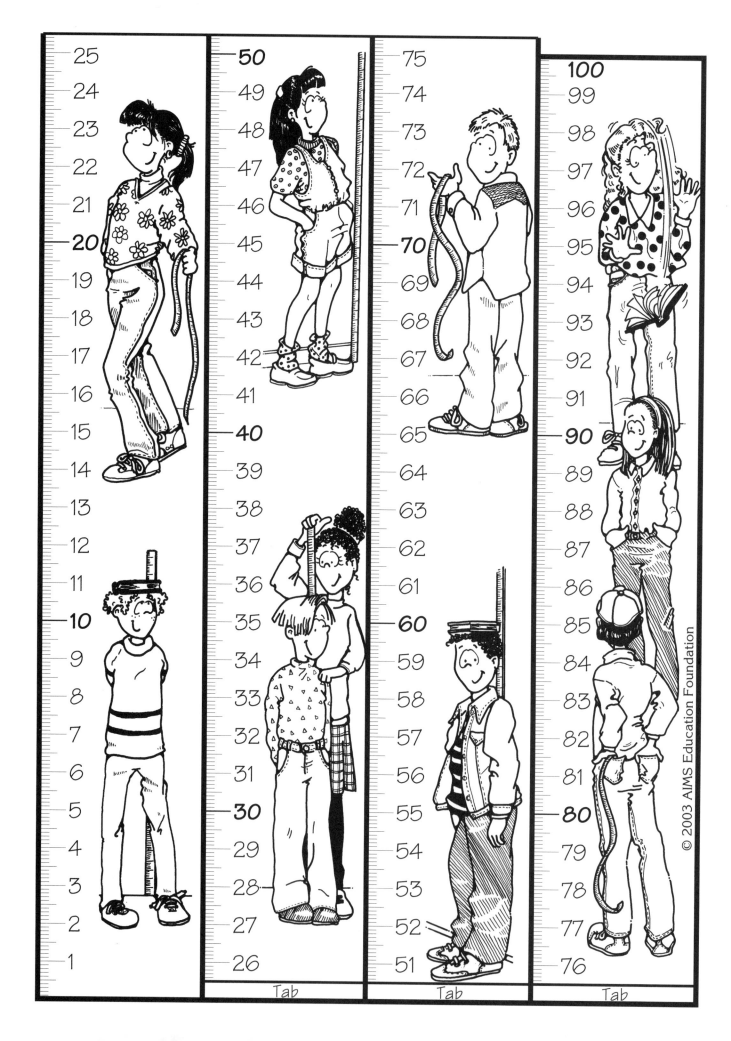

Tab

Tab

Tab

SELECTIVE SERVICE

CONNECTING LEARNING

1. What evidence did you have that a chemical reaction took place?

2. What was the result of this chemical reaction?

3. When you rolled the decalcified egg across your arm and across the paper towel, what did you notice? Why do you think this happened?

4. Explain what you observed when you put the decalcified egg in the corn syrup. ...in the water.

5. Why did this happen?

6. What are some examples of selectively-permeable membranes in living things?

7. Explain in your own words how the game of Red Rover is like what happened with the egg.

8. What things passed through the egg's membrane that would be like the students who got the labels *Pass on Through?*

9. What would be something that *Does not Pass Through?*

10. What are you wondering now?

Butter Battle

Topic
Mixtures

Key Question
What evidence is there that whipping cream is a mixture?

Learning Goals
Students will:
1. identify whipping cream as a specialized mixture called an emulsion, and
2. explore variables that could affect the production of butter.

Guiding Documents
Project 2061 Benchmarks
- *Results of scientific investigations are seldom exactly the same, but if the differences are large, it is important to try to figure out why. One reason for following directions carefully and for keeping records on one's work is to provide information on what might have caused the differences.*
- *Scientific investigations may take many different forms, including observing what things are like or what is happening somewhere, collecting specimens for analysis, and doing experiments. Investigations can focus on physical, biological, and social questions.*

NRC Standards
- *Identify questions that can be answered through scientific investigations.*
- *Design and conduct a scientific investigation.*

NCTM Standard 2000*
- *Recognize and generate equivalent forms of commonly used fractions, decimals, and percents*

Math
Percentages
Ratios

Science
Physical science
  matter
    chemistry

Integrated Processes
Observing
Comparing and contrasting
Communicating
Investigating
Predicting
Collecting and recording data
Interpreting data
Inferring

Materials
For each student group:
  whipping cream
  four baby food jars
  four clean marbles

Background Information
Mayonnaise is an example of an emulsion. Whipping cream is another example. An emulsion is a mixture of at least two normally non-mixable materials. Small particles of one substance are suspended in the particles of another substance. When making butter, the churning process forces water and other non-fat molecules out of the cream so that the emulsion is broken and the fat (butter) congeals and the non-fat components separate. Clumps of globules begin to associate with air bubbles so that a network of air bubbles and fat clumps and globules form entrapping all the liquid and producing a stable foam. If the churning continues, the fat clumps increase in size until they become too large and too few to enclose the air cells, therefore the air bubbles coalesce, the foam begins to "leak" and ultimately butter and butter milk remain.

Management
1. This activity is written so that each student will make his or her own butter in a baby food jar, but the amount of cream will vary in each jar.
2. Student groups of four work best for this activity.
3. All the baby food jars and marbles will need to be the same. The marble acts as the churn in this activity.

Procedure
1. Ask the *Key Question* and state the *Learning Goals*.
2. Assign each student in the group the percentage of the baby food jar that is to be filled with whipping

cream—25%, 50%, 75%, 100%. Discuss the ratios of the filled jars. [Twenty-five percent whipping cream with 75% air is a 1 to 3 ratio. Fifty percent whipping cream to 50% air is a 1 to 1 ratio. Seventy-five percent whipping cream to 25% air is a 3 to 1 ratio. One hundred percent whipping cream to 0% air is a 1 to 0 ratio.) Direct students to fill their baby food jars with the percentages they were assigned. Have them discuss within their groups what they think will happen when their jars are shaken.

3. Direct the students to shake the jar for two minutes. Have them carefully open the lid and make observations of the whipping cream.

4. Tell the students to shake the jar for two more minutes. Have them make observations. (Students should begin to see some butter forming in all except the jar that is 100 percent full with the whipping cream.)

5. Direct them to shake the jars for two more minutes and make observations. Ask them to compare each of the containers and identify the one that produced the most butter after the six minutes of shaking.

6. Discuss with the students the information about emulsions.

Connecting Learning

1. What purpose did the marble serve in this activity?
2. Which whipping cream to air ratio created the most butter?
3. What does this tell you is needed in the production of butter?
4. What evidence did you have that whipping cream is a mixture?
5. Why was it important that we used the same size marble and jar in making the butter?
6. What are you wondering now?

Evidence of Learning

1. Listen as students identify what was changed in the investigation and what was kept the same. Make sure they can identify the variables, the ones controlled and the ones that were changed.
2. Listen during the *Connecting Learning* to make sure they can identify what makes whipped cream an emulsion.

* Reprinted with permission from *Principles and Standards for School Mathematics*, 2000 by the National Council of Teachers of Mathematics. All rights reserved.

Butter Battle

What evidence is there that whipping cream is a mixture?

Learning Goals

STUDENTS WILL:

1. identify whipping cream as a specialized mixture called an emulsion, and

2. explore variables that could affect the production of butter.

Butter Battle

Illustrate the percentage of whipping cream in your jar.

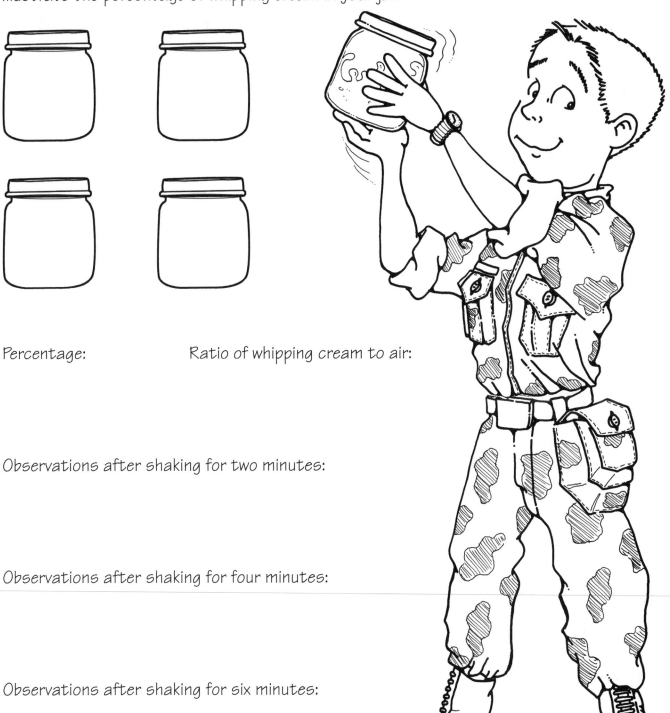

Percentage:

Ratio of whipping cream to air:

Observations after shaking for two minutes:

Observations after shaking for four minutes:

Observations after shaking for six minutes:

Butter Battle

Order the jars by percentage of whipped cream from least amount of butter produced to most amount of butter produced.

What variables remained the same for all jars?

What variable was manipulated?

Describe how you know that whipping cream is a mixture.

Butter Battle

Connecting Learning

1. What purpose did the marble serve in this activity?

2. Which whipping cream to air ratio created the most butter?

3. What does this tell you is needed in the production of butter?

4. What evidence did you have that whipping cream is a mixture?

5. Why was it important that we used the same size marble and jar in making the butter?

6. What are you wondering now?

Topic
Conservation of Mass

Focused Task
Explore the concept of the Conservation of Mass.

Learning Goals
Students will:
1. identify the Law of the Conservation of Mass,
2. demonstrate the Law of the Conservation of Mass through experimentation, and
3. explain the Law of the Conservation of Mass.

Guiding Documents
Project 2061 Benchmarks
- *No matter how parts of an object are assembled, the weight of the whole object made is always the same as the sum of the parts; and when a thing is broken into parts, the parts have the same total weight as the original thing.*
- *Scientists do not pay much attention to claims about how something they know about works unless the claims are backed up with evidence that can be confirmed and with a logical argument.*

NRC Standards
- *Substances react chemically in characteristic ways with other substances to form new substances (compounds) with different characteristic properties. In chemical reactions, the total mass is conserved. Substances often are placed in categories or groups if they react in similar ways; metals is an example of such a group*
- *Develop descriptions, explanations, predictions, and models using evidence.*

NCTM Standard 2000*
- *Understand such attributes as length, area, weight, volume, and size of angle and select the appropriate type of unit for measuring each attribute*

Math
Measurement
   mass

Science
Physical science
   matter

Integrated Processes
Observing
Comparing and classifying
Communicating
Collecting and recording data
Inferring
Generalizing

Materials
For each group of four students:
   balance
   masses
   2 nine-inch round balloons
   2 plastic drink bottles
   4 seltzer tablets
   100-mL graduated cylinder

Background Information
The Law of Conservation of Mass (or Matter) in a chemical reaction can be stated that: *In a chemical reaction, matter is neither created nor destroyed.*

The Law of Conservation of Mass tells us that the mass of the reactants in any chemical reaction equals the mass of the products. Sometimes the products of a chemical reaction look different from the reactants. If a gas is produced, you may not see it. Yet, its mass must be taken into account in finding the total mass of the products. In this activity, the reactants are the seltzer tablets and water. The gas that is produced is the product. Students will be able to observe the gas forming in the water as well as see evidence of the gas as the balloon on the bottle inflates.

The students will observe the reactions twice—once with the balloon on top of the bottle and once with the balloon off the top of the bottle. When the balloon is on the top of the bottle, it traps the gas that is escaping. When the balloon is not on the top of the bottle to trap the gas, there will be a loss of approximately one gram of mass. The students should be able to see this difference and conclude that whether the gas is trapped or it escapes, it is a part of the product and needs to be accounted for in the mass of the product.

The Law of Conservation of Mass is an important law of chemistry. This law holds true for all chemical reactions. Antoine Laurent Lavoisier is given credit for the discovery of this fundamental law in chemistry.

Management

1. Generic seltzer tablets seem to work the best in this investigation.
2. Use nine-inch round balloons.
3. Stress the need to accuracy in all parts of this investigation.

Procedure

1. State the *Focused Task* and the *Learning Goals*.
2. Discuss with the students the Law of the Conservation of Mass. Tell the students that they will be conducting a test that will give evidence that this is true.
3. Distribute the materials to each student group. Tell them to pour 100 mL of water into the bottle.
4. Have the students break the seltzer tablets into small pieces and put the pieces into the balloon.
5. Explain to the students that they must find the initial mass of the bottle with the 100 mL of water and the balloon that has the two broken seltzer tablets in it. Have them record the mass on the student page.
6. Direct the students to pour the seltzer tablets into the bottle while the bottle is still on the balance pan. Tell them to place the balloon back on the balance beside the bottle.
7. Encourage them to make careful observations of what is taking place in the bottle. Have them find the mass after five minutes. (Students should be able to see about a one-gram loss in mass.)
8. Direct a discussion on what they think has happened to the mass since matter cannot be destroyed. [The solid seltzer tablet has undergone a chemical reaction. A gas was released and a temperature change has occurred. The bottle should now feel cool to the touch. Some of the solid mass now is in the form of a gas and that gas has escaped into the air.]
9. Ask the students how could they use the balloon to prove a gas was created as well as to capture the gas? (If no one suggests it, direct the students to place the balloon over the opening of the bottle to capture the gas that is produced from the chemical reaction.
10. Have the students repeat procedures 4–7. At the end of procedure 7 they should find little, if any, loss of mass.

Connecting Learning

1. What is the Law of the Conservation of Mass?
2. How did this investigation help you better understand the Conservation of Mass?
3. Why was it important to include the mass of the balloon both times?
4. How could you explain the Conservation of Mass to someone else based on what you learned in this investigation?
5. What do you think would happen to the mass of a can of soda that is opened and allowed to set out over night? [There will be a loss of mass as a result of the carbon dioxide escaping into the surrounding air. There will also be some loss of mass due to evaporation.]
6. What are you wondering now?

Evidence of Learning

1. Listen as students state the Law of the Conservation of Mass. Make sure that they relate that things can change from one form to another but no mass is lost.
2. Watch as the students are conducting the investigation. Check that they accurately conduct and record their findings. Student work should reflect no loss of mass when the balloon is on the bottle and about one gram loss of mass when the balloon is not on the bottle.

* Reprinted with permission from *Principles and Standards for School Mathematics*, 2000 by the National Council of Teachers of Mathematics. All rights reserved.

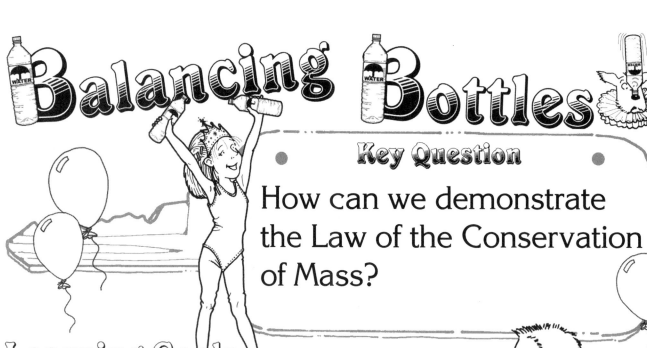

Balancing Bottles

How can we demonstrate the Law of the Conservation of Mass?

Learning Goals

STUDENTS WILL:

1. identify the Law of the Conservation of Mass,

2. demonstrate the Law of the Conservation of Mass through experimentation, and

3. explain the Law of the Conservation of Mass.

Balancing Bottles

add 100 mL
of water

put
bottle,
seltzer
tablets,
and the
balloon
in the
balance

find
the
mass

Initial mass:
100 mL of water, bottle,
balloon, 2 seltzer tablets

_____

Now add the
seltzer tablets
to the bottle
of water. Wait
five minutes.

Mass after reaction

_____

What happened?

Why do you think this happened?

  In words and pictures, show how you are going to revise this
experiment to capture the gas that escaped. Be sure to include
the masses of the objects before and after the reaction.

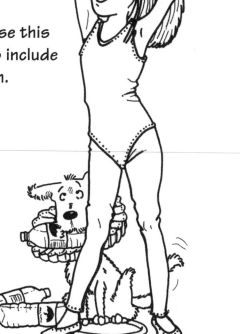

148

Balancing Bottles

If mass cannot be destroyed, what happened here?

Write in your own words what the Law of the Conservation of Mass means.

How were you able to prove that matter was conserved?

Connecting Learning

1. What is the Law of the Conservation of Mass?

2. How did this investigation help you better understand the Conservation of Mass?

3. Why was it important to include the mass of the balloon both times?

4. How could you explain the Conservation of Mass to someone else based on what you learned in this investigation?

5. What do you think would happen to the mass of a can of soda that is opened and allowed to set out over night?

6. What are you wondering now?

Curds and Weigh

Topic
Conservation of Mass

Key Question
How does the mass of the materials before a chemical reaction compare with the mass of the materials after the chemical reaction is complete?

Learning Goal
Students will observe that matter is conserved even when it goes through a chemical reaction.

Guiding Documents
Project 2061 Benchmarks
- *No matter how many parts of an object are assembled, the weight of the whole object made is always the same as the sum of the parts and when a thing is broken into parts, the parts will have the same total weight as the original thing.*
- *When a new material is made by combining two or more chemicals, it has properties that are different from the original materials. For that reason, a lot of different materials can be made from a small number of basic kinds of materials.*
- *No matter how substances within a closed system interact with one another or how they combine or break apart, the total weight of the system remains the same. The idea of atoms explains the conservation of matter. The number of atoms stays the same no matter how they are rearranged thus their total mass stays the same.*

NCTM Standards 2000*
- *Understand such attributes as length, area, weight, volume, and size of angle and select the appropriate type of unit for measuring each attribute*
- *Recognize and apply mathematics in contexts outside of mathematics*

Math
Measurement
   mass

Science
Physical science
   matter

Integrated Processes
Observing
Predicting
Collecting and recording data
Interpreting data
Generalizing
Applying

Materials
For each group:
   150 mL (5/8 cup) of milk (whole or 2%)
   thermometer that reads to at least 80° C (176° F)
   hot plate
   sauce pan or beaker
   two tablets of rennin (see *Management 1*)
   stirring rod
   clock, watch, or timer

Background Information
When milk separates into curds and whey, we say that it has curdled or coagulated. Remember when little Miss Muffet sat on her tuffet and ate her curds and whey? She was really eating curdled milk which is much like cottage cheese, only wetter and, if it is made with an acid, more sour (see *Extension 1*).

Casein is the principle protein compound of milk. In milk, casein is not in solution, but rather in a suspension. When rennin is added to milk, the casein is transformed into a new compound called paracasein. There is the suggestion that a chemical change has occurred in this reaction since the process to produce paracasein has never been reversed to produce casein.

Rennin, like all enzymes, has an optimum temperature range. It is active from about 10° C to about 80° C. Below 10° C, the rennin is inactive and above 80° C, it is denatured, that is to say the enzyme is changed into something new and therefore no longer causes milk to form curds.

The curd contains most of the fat, casein protein, and vitamin A of the original milk. Whey is about 93% water, but contains proteins, some minerals, vitamins, and most of the lactose of the original milk. In North America, the leftover whey is sold as a nutritious additive for bread, ice cream, processed luncheon meats, and even food for animals.

In this activity students will find the mass of the arrangements for two identical setups consisting of milk, rennin tablets, beakers, and stir rods. They will then add a rennin tablet to one recipe and allow the other to remain as their control. After the chemical reaction has occurred, the students will compare the masses of the two recipes and discover that they remained the same. The matter was conserved.

The famous scientist Antoine-Laurent Lavoisier is credited with the discovery that the mass of the original materials and the products of a chemical reaction remain the same. No mass is lost during a chemical reaction. He uncovered a very important principle, the *conservation of mass: Nothing is created or destroyed, only alternations and modifications; there is an equal amount of matter before and after the operation.*

Management
1. Rennin is available under the trademark Junket®, and is used in making ice cream, jelly, refrigerated pies, and cheese.
2. If you are going to allow your students to taste the curds, avoid potential contamination by making sure all utensils and containers have not been used for any other experiments.
3. Because of the potential breakage of thermometers, alcohol thermometers should be used. Caution students not to touch the hot plate while it is hot.
4. Heat-proof containers can be used instead of beakers. Also, metal spoons can be used for stir rods. Make certain that the containers and spoons have identical masses.

Procedure
1. Prepare two beakers or heat-proof containers of milk with 75 mL of milk in each one.

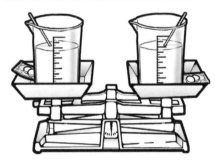

2. Place one beaker on either side of a balance. Put a stir rod in each beaker and a rennet tablet alongside each beaker. The balance should equalize. Compensate by adding mass to one pan until it does. Have the students record their observations.
3. Warm both beakers on a hot plate to a temperature of 30° to 40°C. Do not boil.
4. Add one tablet of rennin to one of the beakers and stir for several minutes. Observe that the milk forms curds. Have students make a drawing of the before and after beakers.
5. Using a balance, place one of your beakers on either side. Make sure both beakers have a stirring

rod in them. To insure that both sides of the balance have the same components for which you are determining the mass, make certain that the rennin tablet is placed beside the beaker containing milk that has only been heated.
6. Direct the students to observe as to whether the mass on each side of the balance is the same (i.e., Does the balance equalize?). Have students record their observations.

Connecting Learning
1. What did you discover as to the mass of the materials before and after the investigation?
2. What could you say about the mass of the original material and the resulting mass of the products of a chemical reaction?
3. How do the products of the reaction differ from the original ingredients? Does this reveal anything about what happens during a chemical reaction?
4. What are some possible indicators that a chemical reaction has taken place?
5. What are you wondering now?

Extensions
1. Try producing the curds and whey with the addition of an acid such as vinegar, pineapple juice, or grape juice instead of the rennin. (FYI—The reaction is slightly different with acids. Acids release casein from combination with calcium and dissolve calcium phosphate, whereupon free casein precipitates leaving the calcium salts in the whey. Whew!)
2. If you want to eat the curds, put the mixture in a bowl and refrigerate it. When the curds and whey are cool, they will be ready to eat. If you like, add salt or sugar to taste.

* Reprinted with permission from *Principles and Standards for School Mathematics,* 2000 by the National Council of Teachers of Mathematics. All rights reserved.

Curds and Weigh

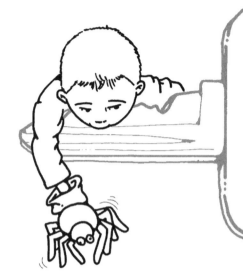

How does the mass of the materials before a chemical reaction compare with the mass of the materials after the chemical reaction is complete?

Learning Goal

STUDENTS WILL:

observe that matter is conserved even when it goes through a chemical reaction.

Curds and Weigh

Show your observations in words and pictures.

Before the reaction

After the reaction

Curds and Weigh

In words and pictures, show your conclusions about the masses before and after the reaction.

Curds and Weigh

Connecting Learning

1. What did you discover as to the mass of the materials before and after the investigation?

2. What could you say about the mass of the original material and the resulting mass of the products of a chemical reaction?

3. How do the products of the reaction differ from the original ingredients? Does this reveal anything about what happens during a chemical reaction?

4. What are some possible indicators that a chemical reaction has taken place?

5. What are you wondering now?

BLOCK BUSTERS

Topic
Conservation of Matter

Key Questions
1. How can you model how plants get energy from the sun?
2. How do living things turn sugar molecules into energy?

Learning Goals
Students will:
1. identify photosynthesis as the process of turning the sun's energy into sugar,
2. identify how living things convert sugar into energy for life's processes, and
3. model the Law of the Conservation of Mass.

Guiding Documents
Project 2061 Benchmarks
- *No matter how parts of an object are assembled, the weight of the whole object made is always the same as the sum of the parts; and when a thing is broken into parts, the parts have the same total weight as the original thing.*
- *Materials may be composed of parts that are too small to be seen without magnification.*
- *When a new material is made by combining two or more materials, it has properties that are different from the original materials. For that reason, a lot of different materials can be made from a small number of basic kinds of materials.*

NRC Standard
- *Different kinds of questions suggest different kinds of scientific investigations. Some investigations involve observing and describing objects, organisms, or events; some involve collecting specimens; some involve experiments; some involve seeking more information; some involve discovery of new objects and phenomena; and some involve making models.*

Science
Physical science
    matter
Life science
    photosynthesis
    transpiration

Integrated Processes
Observing
Relating
Communicating
Inferring
Generalizing
Applying

Materials
For each student:
    student booklet
    hex-a-link cubes in different colors

For each student group:
    balance
    masses
    colored pencils

Background Information
The Law of Conservation of Mass (or Matter) in a chemical reaction is: *In a chemical reaction, matter is neither created nor destroyed.* The Law of Conservation of Mass tells us then that the mass of the reactants in any chemical reaction equals the mass of the products. Sometimes the products of a chemical reaction look different from the reactants.

This experience will examine two exchange processes that take place in plants and animals.
1. *Photosynthesis* is the process by which plants convert energy from the sun into sugar and oxygen. This occurs in the chloroplasts of plants. Energy from the sun plus carbon dioxide and water are converted into sugar and oxygen. The chemical equation is:

Energy + 6 CO_2 + 6 H_2O changed into $C_6H_{12}O_6$ + 6 O_2.

2. *Cellular respiration* is the process by which living things break down the sugar molecules to produce energy. This process is just the opposite of photosynthesis. This chemical equation looks like this:

$C_6H_{12}O_6$ + 6 O_2 changed into 6 CO_2 + 6 H_2O.

This experience will allow the students to build a model of this fundamental life process as well as identify how matter is conserved.

The Law of Conservation of Mass is an important law of chemistry. This law holds true for all chemical reactions. Antoine Laurent Lavoisier is given credit for the discovery of this fundamental law in chemistry.

Management

1. Each student will need three different colored hex-a-link cubes; six in one color to represent carbon, 12 in the second color to represent hydrogen, and 18 in the third color to represent oxygen. The students will need to create a color key in their booklet based on the colors they are using.

Procedure

1. Read the *Key Questions* and state the *Learning Goals*.
2. Distribute the student books and hex-a-link cubes to the students. Read and discuss the first page together.
3. Ask the students to read the booklet and construct the different molecules as described in the reading passages.

4. Direct a discussion on photosynthesis and cellular respiration.
5. Have the students find the total mass of the cubes from each of the equations.

Connecting Learning

1. How does the Law of the Conservation of Mass relate to this activity?
2. How did the models help you better understand the processes of cellular respiration and photosynthesis?
3. What do the processes of cellular respiration and photosynthesis have in common? How are the two different?
4. Why do you think scientists use chemical formulas to describe things?
5. What molecule is larger, a water molecule or a sugar molecule? How do you know?
6. What did the balance show us in this activity?
7. What are you wondering now?

BLOCK BUSTERS

Key Questions

1. How can you model how plants get energy from the sun?
2. How do living things turn sugar molecules into energy?

Learning Goals

STUDENTS WILL:

1. identify photosynthesis as the process of turning the sun's energy into sugar,

2. identify how living things convert sugar into energy for life's processes, and

3. model the Law of the Conservation of Mass.

CO_2

H_2O

This change releases energy that the cell can now use. Rearrange the hex-a-links to make the six carbon dioxide molecules and the six water molecules. This is an ongoing process that passes energy from the sun into living things.

This chemical equation looks like this:

$$C_6H_{12}O_6 + 6O_2 \longrightarrow 6CO_2 + 6H_2O.$$

$C_6H_{12}O_6$

HYDROGEN

OXYGEN

CARBON

O_2

CO_2

H_2O

Photosynthesis and Cellular Respiration

Two very important exchange processes connect living things on Earth. These processes are photosynthesis and cellular respiration.

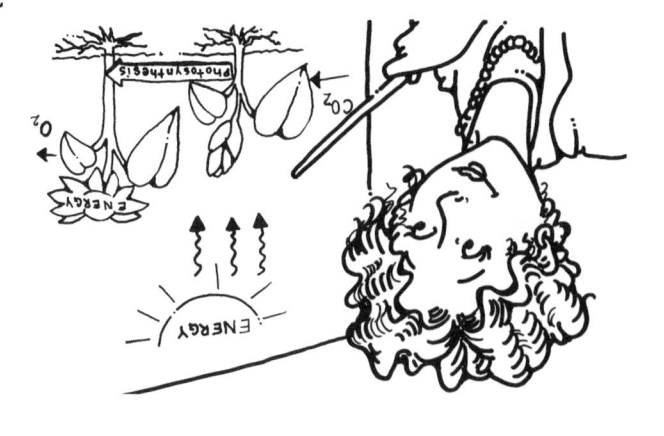

- Photosynthesis is the way plants are able to turn energy from the sun into sugar molecules. Sugar is a term used to describe food that all living things need.

Cellular respiration takes the sugar molecule and the six oxygen molecules. It changes them into six carbon dioxide molecules and six molecules of water.

- Cellular respiration is a process that takes place in the cells of living things. The cells break the sugar molecules down and turn it into food.

You will need to break the models of the carbon dioxide and water apart from the first part of the formula in order to make the six oxygen molecules in the second part. All of the remaining hex-a-link cubes are the atoms in a single molecule of sugar.

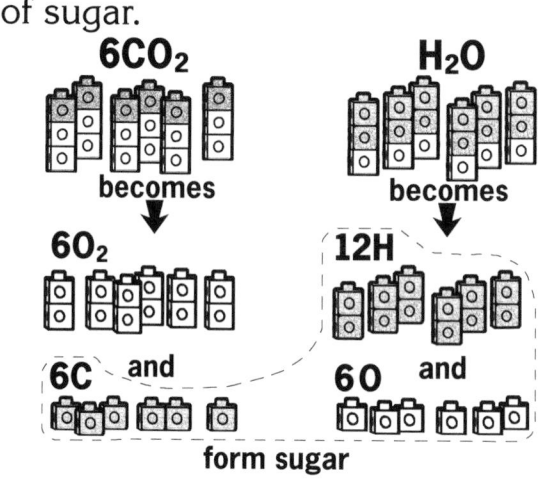

One of the important ideas in science is that matter, which is what molecules are, cannot be created or destroyed, but it can be changed. These molecules you have built can be changed into a sugar molecule and oxygen molecules.

Building Molecule Models

Scientists use chemical formulas to tell what elements are in different molecules. Carbon dioxide's chemical formula is CO_2. That means that there is a total of three atoms in a carbon dioxide molecule, one of carbon and two of oxygen.

The second part of the equation for photosynthesis is $C_6H_{12}O_6 + 6\ O_2$. The first molecule is the sugar and the second is the oxygen. The formulas tell you that each of the six oxygen molecules is made up of two oxygen atoms.

The sun is the source that plants use to begin the process of photosynthesis. Plants have special parts in their cells called chloroplasts. This is the place where photosynthesis occurs. Plants need the energy from the sun in the form of light plus two other "ingredients," carbon dioxide and water.

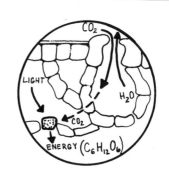

11

4

To build a model, select one color of the hex-a link cubes to represent the element carbon and one color to represent the element oxygen. Connect the three cubes to make a model of a carbon dioxide molecule. You will need to build five more carbon dioxide molecules.

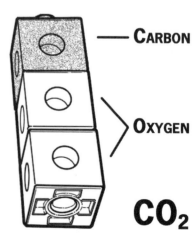

CARBON

OXYGEN

CO_2

7

You now need to construct a model of a water molecule. Water's chemical formula is H_2O. That tells you that there are two atoms of hydrogen and one atom of oxygen in a water molecule.

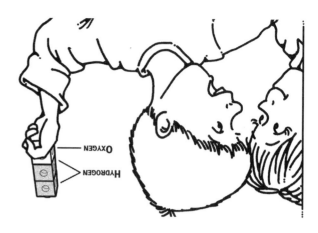

OXYGEN

HYDROGEN

H_2O

OXYGEN

HYDROGEN

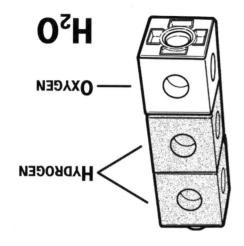

You will use the same color hex-a-link cube you used for the oxygen in the carbon dioxide molecule and the third color for the hydrogen atom. You will need to build a total of six water molecules.

9

You should have 12 molecule models now, six carbon dioxide and six water. The chemical equation of what you have built looks like this:

Energy + $6CO_2$ + $6H_2O$

CO_2 H_2O

CONNECTING LEARNING

1. How does the Law of the Conservation of Mass relate to this activity?

2. How did the models help you better understand the processes of cellular respiration and photosynthesis?

3. What do the processes of cellular respiration and photosynthesis have in common? How are the two different?

4. Why do you think scientists use chemical formulas to describe things?

5. What molecule is larger, a water molecule or a sugar molecule? How do you know?

6. What did the balance show us in this activity?

7. What are you wondering now?

Topic
Reactions in Matter

Key Question
What can you learn about a substance by adding cabbage juice to it?

Learning Goals
Students will:
1. identify that a color change is an indication that a chemical reaction has occurred;
2. read a chart to approximate a substance's pH level; and
3. classify a substance as an acid, base, or neutral as determined by its pH.

Guiding Documents
Project 2061 Benchmark
* *Results of scientific investigations are seldom exactly the same, but if the differences are large, it is important to try to figure out why. One reason for following directions carefully and for keeping records on one's work is to provide information on what might have caused the differences.*

NRC Standard
* *Substances react chemically in characteristic ways with other substances to form new substances (compounds) with different characteristic properties. In chemical reactions, the total mass is conserved. Substances often are placed in categories or groups if they react in similar ways; metals are an example of such a group.*

Science
Physical science
   chemistry
      chemical reactions
      pH levels

Integrated Processes
Observing
Comparing and classifying
Communicating
Collecting and recording data
Interpreting data

Materials
For the class:
   lime solution (see *Management 3*)
   washing soda solution (see *Management 3*)
   baking soda solution (see *Management 3*)
   cream of tartar solution (see *Management 3*)
   lemon juice
   household cleaning solution
   ammonia
   distilled water
   tap water
   milk of magnesia
   vinegar
   milk
   12 eyedroppers
   12 3-ounce cups
   12 *Solution Station Cards* (see *Management 4*)

For each student group:
   cabbage juice indicator (see *Management 1*)
   12 1-ounce portion cups
   eyedropper
   solutions chart

Background Information
An indicator is a substance that changes color as an indication that two or more substances have reacted. In this activity, red cabbage leaves and water are used to prepare an indicator. The students will approximate the pH level of the substance by reading a simplified pH chart. pH is used by chemist to indicate the concentration of hydrogen ions in a solution. The letters *pH* stand for *potential of hydrogen*. Chemists use indicators to test whether a substance is an acid or a base. Indicators work by turning a distinctive color in the presence of an acid or a base.

Management
1. Chop red cabbage and place in a dish that can be microwaved. Add three to four cups of water and cover the dish. Microwave on high for about seven minutes. Leave the lid on the dish and let the ingredients cool. When cooled, drain off the cabbage juice. The liquid should be a dark reddish purple color. Each group will need a 3-oz. cup filled approximately two-thirds full of cabbage juice.
2. Cabbage juice can be stored in the refrigerator for several days or for an extended period in the freezer.

3. The powdered substances need to be in solution. Mix the washing soda, the baking soda, the pickling lime, and the cream of tarter with distilled water. Use a three-to-one ratio when mixing the solutions (three parts water to one part powder). Some of the powder may not completely dissolve.

4. You will need to designate a place in the room that will be called the *Solution Station*. Students will bring the 1-ounce cups to this area to get the solution to test. You will need to prepare 12 *Solution Station Cards*. Record the name of a different solution on each card. Laminate the cards for durability. Place the appropriate cup of solution on each card. Remind the students they will need to make sure they bring a new cup each time and tell them the eyedropper used for the solution must be placed back on the solution station card beside the solution.

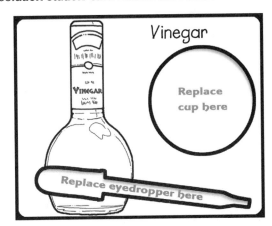

5. The acids will be the lemon juice, vinegar, cream of tartar (tartaric acid). Distilled water is the only substance that should be neutral. Tap water may be slightly basic. Baking soda is also a weak base. The strong bases will be the household cleaners, ammonia, milk, washing soda, milk of magnesia, and the pickling lime. No base will have a pH of 13 or 14.

6. Warn the students not to taste any of the substances they will be testing.

7. Groups of three to four work well for this activity.

Procedure

1. Ask the *Key Question* and state the *Learning Goals*.

2. Ask the students if they have ever heard the term pH. Tell them pH tells us if a substance is an acid or a base or somewhere between. Review with them the information on the student page about acids and bases. Remind them that a change in color is an indication that a chemical reaction has occurred.

3. Distribute the cabbage juice indicator, an eyedropper, and 12 portion cups to each group.

4. Request that one student from each group come to the solution table area and place two or three droppers full of the lime water into one of the portion cups. Have them return to their groups.

5. Guide students in dropping three of four drops of the cabbage juice into the lime water solution.

6. Direct them to use the color chart to estimate pH. Place the cup in the first block on the solutions page. Have them use a pencil to record the estimated pH and color so that they can adjust the estimate after they have tested and compared all the substances.

7. Ask students to test the other 11 substances in the same manner, testing one at a time so as to keep the names of the substances correct. Point out that the order in which they test the substances does not matter.

8. Tell the students to sequence the solutions based on their determination of pH from most acidic to most basic.

9. Tell the students to identify each substance as an acid, a base, or neutral.

Connecting Learning

1. Which substances were acids? Which were bases?

2. Farmers sometimes add lime to soil if the soil is too acidic. What would the lime do to the soil? Explain why you think this.

3. In what order did your group rank the solutions? Did all the groups rank the solutions in the same order? Why do you think you got the results you did?

4. Why do you think distilled water has a different pH than tap water? [Distilled water has been conditioned to remove impurities that will react to the indicator solution.]

5. In which solutions are you sure that a chemical reaction took place? How do you know? [The ones that had a clear color change take place.]

Evidence of Learning

1. Check student sheets for accuracy in recording pH for each substance and correct identification of acid, base, or neutral for the substance.

2. Listen for student response in answering the *Connecting Learning* questions.

* Reprinted with permission from *Principles and Standards for School Mathematics*, 2000 by the National Council of Teachers of Mathematics. All right reserved.

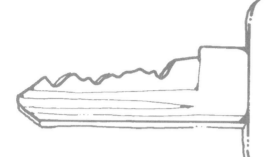

Key Question

What can you learn about a substance by adding cabbage juice to it?

Learning Goals

The students will:

1. identify that a color change is an indication that a chemical reaction has occurred;

2. read a chart to approximate a substance's pH level; and

3. classify a substance as an acid, base, or neutral as determined by its pH.

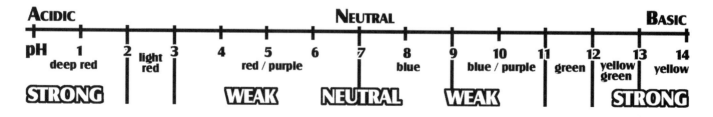

Station Cards

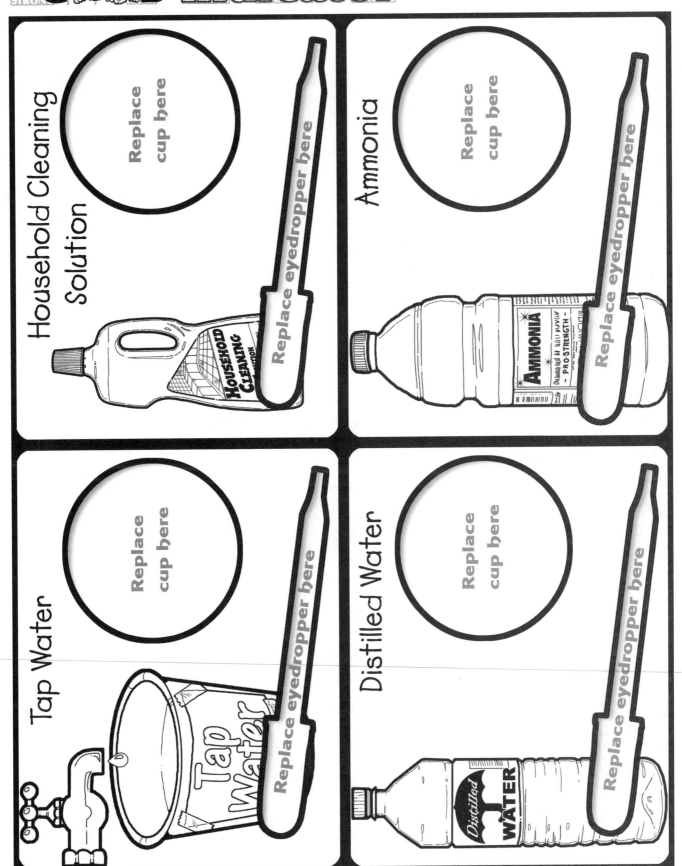

An indicator is a substance that changes color as an indication that two or more substances have reacted. Indicators work by turning a distinctive color in the presence of an acid or a base. In this activity, cabbage leaves and water are used to make an indicator. Chemists use pH levels to determine whether a substance is an acid or a base. You will determine the approximate pH level of the substance by reading the simplified pH chart.

ACIDIC   **NEUTRAL**   **BASIC**

| pH | 1 deep red | 2 light red | 3 | 4 | 5 red / purple | 6 | 7 | 8 blue | 9 | 10 blue / purple | 11 | 12 green | 13 yellow green | 14 yellow |

STRONG   **WEAK**   **NEUTRAL**   **WEAK**   **STRONG**

Solution	Color	Estimated pH	Acid, Base, or Neutral
lime			
washing soda			
baking soda			
cream of tartar			
lemon juice			
household cleaning			
ammonia			
distilled water			
tap water			
milk of magnesia			
vinegar			
milk			

most acidic _____   _____   _____

_____   _____   _____

_____   _____   _____

_____   _____   _____ most basic

Connecting Learning

1. Which substances were acids? Which were bases?

2. Farmers sometimes add lime to soil if the soil is too acidic. What would the lime do to the soil?

3. In what order did your group rank the solutions? Were all the groups the same? Why do you think you got the results you did?

4. Why do you think distilled water has a different pH than tap water?

5. In which solutions are you sure a chemical reaction took place? Why?

6. What are you wondering now?

Mixed Reactions

Topic
Chemical Reactions

Key Question
What are some of the ways you can tell if a chemical reaction has taken place?

Learning Goals
Students will:
1. conduct tests to check for a chemical reaction,
2. classify the tests based on whether or not a chemical reaction has taken place, and
3. identify the production of gas and change of temperature as evidence of a chemical reaction.

Guiding Documents
Project 2061 Benchmark
• When a new material is made by combining two or more materials, it has properties that are different from the original materials. For that reason, a lot of different materials can be made from a small number of basic kinds of materials.

NRC Standard
• Substances react chemically in characteristic ways with other substances to form, new substances (compounds) with different characteristic properties. In chemical reactions, the total mass is conserved. Substances often are placed in categories or groups if they react in similar ways; metal is an example of such a group.

NCTM Standards 2000*
• Collect data using observations, surveys, and experiments
• Recognize the differences in representing categorical and numerical data

Math
Data analysis

Science
Physical science
    chemical reactions

Integrated Processes
Observing
Comparing and contrasting
Classifying
Collecting and recording data
Interpreting data
Inferring

Materials
For the whole class:
    clock or watch
    hydrogen peroxide
    calcium chloride (see *Management 7*)
    baking soda
    vinegar
    sugar
    salt
    six buckets or tubs
    six plastic trash cans (see *Management 4*)
    paper towels
    overhead transparency of *Evidence of a Chemical Reaction* page
    station cards

For each group:
    one clear plastic cup, 9-oz. (see *Management 5*)
    one plastic spoon
    one thermometer

For each student:
    data sheet

Background Information
A chemical reaction is another name for a chemical change. A chemical change is the process that takes place when one substance turns into another substance. For example, when an iron nail rusts, it undergoes a chemical change. The iron in the nail reacts with the oxygen in the air producing a new substance, iron oxide, which is rust.

A change in temperature is one piece of evidence the students can use to see if a chemical reaction has taken place. Reactions can be endothermic or exothermic. In an endothermic reaction, the temperature will decrease. In an exothermic reaction, the temperature will increase. A second piece of evidence that a chemical reaction has taken place is the production of a gas (bubbles). A third piece of evidence is the formation of a precipitate. A precipitate is a solid substance that forms within a liquid (or even in some gases). Other indications of a chemical reaction are production of light and change of color.

In this activity, students will examine six different reactions. They will identify whether a chemical reaction has taken place by looking specifically for temperature changes and the production of gas.

Management

1. Caution the students not to taste any of the substances at the stations. Tell the students it is not safe to mix unknown substances to test for reactions. The materials they will be testing are known not to produce harmful reactions.
2. Prepare the station cards. Make an overhead transparency of the *Evidence of a Chemical Reaction* sheet.
3. You can make a plastic trash can to use at each of the stations by cutting the top four inches off a two- or three-liter drink bottle.
4. Tell the students that they must clean their equipment thoroughly between each station. Place a bucket or tub at each station with water in it so that the students can clean the cup, spoon, and thermometer.

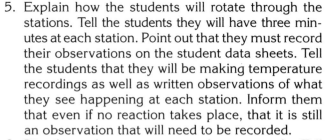

5. You will need to attach the graduated scale slips to the sides of the 9-ounce clear cups. The strips need to be completely covered with clear tape so that they will be waterproof.
6. Student groups of four work best for this activity.
7. Calcium chloride can be purchased at hardware stores and grocery stores. It is used to thaw ice on sidewalks and driveways.

Tape graduated scale slip to the outside facing in.

Cover completely with tape.

Procedure

1. Ask the *Key Question* and state the *Learning Goals*.
2. Tell the students that they will rotate through six stations. Explain that they will be observing how selected solids react to selected liquids. Discuss the indicators on the *Evidence of a Chemical Reaction* sheet.
3. Distribute a cup, spoon, and thermometer to each group. Also hand out a data sheet to each student. Show how to fold the data sheet to form a booklet.

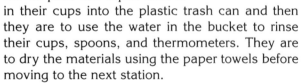

4. Direct the students' attention to the bucket and plastic trash can at each station. Tell them at the end of each test, they are to pour the liquid that is in their cups into the plastic trash can and then they are to use the water in the bucket to rinse their cups, spoons, and thermometers. They are to dry the materials using the paper towels before moving to the next station.

5. Explain how the students will rotate through the stations. Tell the students they will have three minutes at each station. Point out that they must record their observations on the student data sheets. Tell the students that they will be making temperature recordings as well as written observations of what they see happening at each station. Inform them that even if no reaction takes place, that it is still an observation that will need to be recorded.
6. Distribute the temperature change graph. Tell students to record the initial temperature for each station. Have them create a scale for the graph based on the data collected and graph the temperature changes. Increases in temperature will be above the initial temperatures and decreases will be below. If no temperature change occurred, have students record that by writing "0" in the spaces provided to record temperature change.

Connecting Learning

1. What evidence do you have as to whether or not a chemical reaction took place at each station?
2. How are the temperature changes you observed in these investigations different from those that would happen if you put something on the stove or in a refrigerator?
3. If you pour hydrogen peroxide on a cut and a white substance appears, what do you know has taken place? [a chemical reaction]
4. Why was it important to record what you observed at each station?
5. How did the thermometer help determine if a chemical reaction had taken place? [A change in temperature is an indication of a chemical reaction.]
6. Did some of the reactions have more than one thing happening that would tell you a chemical reaction has taken place? Explain. (Some will have both a temperature change and production of a gas.)

Evidence of Learning

1. Listen to student discussion during the *Connecting Learning* questions. Students should state evidence from the investigation in describing chemical reactions.
2. Check the student recording sheets to see that they accurately identify the stations that produced a chemical reaction and the evidence that it took place.
   - hydrogen peroxide and activated yeast—temperature change and gas produced
   - hydrogen peroxide and sugar—no reaction
   - water and calcium chloride—temperature change
   - water and baking soda—temperature change
   - vinegar and baking soda—temperature change and gas produced
   - vinegar and salt—no reaction

* Reprinted with permission from *Principles and Standards for School Mathematics*, 2000 by the National Council of Teachers of Mathematics. All rights reserved.

Mixed Reactions

Graduated Scales

Tape graduated scale slip to the outside facing in.

Cover completely with tape.

240 mL	240 mL	240 mL	240 mL	240 mL	240 mL	240 mL	240 mL
220 mL	220 mL	220 mL	220 mL	220 mL	220 mL	220 mL	220 mL
200 mL	200 mL	200 mL	200 mL	200 mL	200 mL	200 mL	200 mL
180 mL	180 mL	180 mL	180 mL	180 mL	180 mL	180 mL	180 mL
160 mL	160 mL	160 mL	160 mL	160 mL	160 mL	160 mL	160 mL
140 mL	140 mL	140 mL	140 mL	140 mL	140 mL	140 mL	140 mL
120 mL	120 mL	120 mL	120 mL	120 mL	120 mL	120 mL	120 mL
100 mL	100 mL	100 mL	100 mL	100 mL	100 mL	100 mL	100 mL
80 mL	80 mL	80 mL	80 mL	80 mL	80 mL	80 mL	80 mL
60 mL	60 mL	60 mL	60 mL	60 mL	60 mL	60 mL	60 mL
40 mL	40 mL	40 mL	40 mL	40 mL	40 mL	40 mL	40 mL
20 mL	20 mL	20 mL	20 mL	20 mL	20 mL	20 mL	20 mL
0 mL	0 mL	0 mL	0 mL	0 mL	0 mL	0 mL	0 mL

240 mL	240 mL	240 mL	240 mL	240 mL	240 mL	240 mL	240 mL
220 mL	220 mL	220 mL	220 mL	220 mL	220 mL	220 mL	220 mL
200 mL	200 mL	200 mL	200 mL	200 mL	200 mL	200 mL	200 mL
180 mL	180 mL	180 mL	180 mL	180 mL	180 mL	180 mL	180 mL
160 mL	160 mL	160 mL	160 mL	160 mL	160 mL	160 mL	160 mL
140 mL	140 mL	140 mL	140 mL	140 mL	140 mL	140 mL	140 mL
120 mL	120 mL	120 mL	120 mL	120 mL	120 mL	120 mL	120 mL
100 mL	100 mL	100 mL	100 mL	100 mL	100 mL	100 mL	100 mL
80 mL	80 mL	80 mL	80 mL	80 mL	80 mL	80 mL	80 mL
60 mL	60 mL	60 mL	60 mL	60 mL	60 mL	60 mL	60 mL
40 mL	40 mL	40 mL	40 mL	40 mL	40 mL	40 mL	40 mL
20 mL	20 mL	20 mL	20 mL	20 mL	20 mL	20 mL	20 mL
0 mL	0 mL	0 mL	0 mL	0 mL	0 mL	0 mL	0 mL

Evidence of a Chemical Reaction

A change in temperature

Gas is produced

Light is produced

A change in color

A precipitate forms

Mixed Reactions

Learning Goals

Students will:

1. conduct tests to check for a chemical reaction,

2. classify the tests based on whether a chemical reaction has taken place, and

3. identify the production of a gas and a temperature change as evidence of a chemical reaction.

STATION 2

Hydrogen Peroxide and Sugar

Temperature

before

after

ending

check
one

□ chemical change   □ no chemical change

Observations

STATION 1

Hydrogen Peroxide and Activated Yeast

Observations

Temperature

before

after

ending

check
one

□ chemical change   □ no chemical change

Mixed Reactions!

STATION 3

Vinegar and Salt

Water and Calcium Chloride

Temperature

before

after

ending

check
one

□ chemical change   □ no chemical change

Observations

STATION 6

Observations

Temperature

before

after

ending

check
one

□ chemical change   □ no chemical change

STATION 4

Vinegar and Baking Soda

Water and Baking Soda

Temperature

before

after

ending

check
one

□ chemical change   □ no chemical change

Observations

STATION 5

Observations

Temperature

before

after

ending

check
one

□ chemical change   □ no chemical change

Mixed Reactions

Graph the temperature change (if there was one) for each reaction. Be sure to write the beginning temperatures at each station in the spaces provided. Label the graph along the left side with the appropriate temperature increments based on your results.

Temperature Change: + ____ °C   + ____ °C   + ____ °C   + ____ °C   + ____ °C   + ____ °C

Temperature Increase in °C

0

Beginning Temperature in °C

____ Station One   ____ Station Two   ____ Station Three   ____ Station Four   ____ Station Five   ____ Station Six

0

Temperature Decrease in °C

Temperature Change: − ____ °C   − ____ °C   − ____ °C   − ____ °C   − ____ °C   − ____ °C

© 2003 AIMS Education Foundation

Hydrogen Peroxide
and Activated Yeast

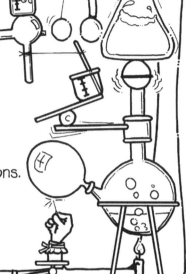

Procedure
1. Pour 40mL of hydrogen peroxide into your cup.
2. Find and record the initial temperature of the hydrogen peroxide.
3. Add one spoonful of activated yeast to the cup and stir.
4. Find and record the temperature after adding the activated yeast.
5. Observe for three minutes and record any observations.
6. Find and record the ending temperature.
7. Clean your equipment and area.

Hydrogen Peroxide
and Sugar

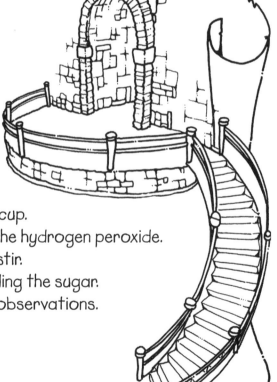

Procedure
1. Pour 40mL of hydrogen peroxide into your cup.
2. Find and record the initial temperature of the hydrogen peroxide.
3. Add one spoonful of sugar to the cup and stir.
4. Find and record the temperature after adding the sugar.
5. Observe for three minutes and record any observations.
6. Find and record the ending temperature.
7. Clean your equipment and area.

Water and Calcium Chloride

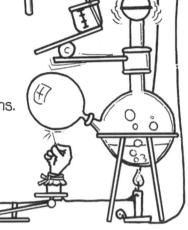

Procedure

1. Pour 40mL of water into your cup.
2. Find and record the initial temperature of the water.
3. Add one spoonful of the calcium chloride to the cup and stir.
4. Find and record the temperature after adding the calcium chloride.
5. Observe for three minutes and record any observations.
6. Find and record the ending temperature.
7. Clean your equipment and area.

Water and Baking Soda

Procedure

1. Pour 40mL of water into your cup.
2. Find and record the initial temperature of the water.
3. Add one spoonful of the baking soda to the cup and stir.
4. Find and record the temperature after adding the baking soda.
5. Observe for three minutes and record any observations.
6. Find and record the ending temperature.
7. Clean your equipment and area.

Vinegar and Baking Soda

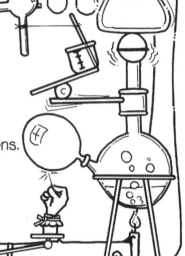

Procedure

1. Pour 40mL of vinegar into your cup.
2. Find and record the initial temperature of the vinegar.
3. Add one spoonful of the baking soda to the cup and stir.
4. Find and record the temperature after adding the baking soda.
5. Observe for three minutes and record any observations.
6. Find and record the ending temperature.
7. Clean your equipment and area.

Vinegar and Salt

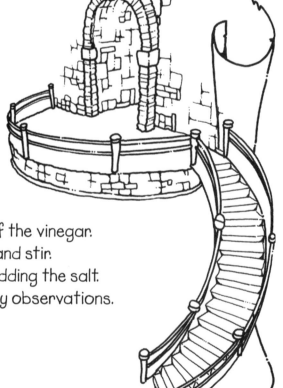

Procedure

1. Pour in 40mL of vinegar into your cup.
2. Find and record the initial temperature of the vinegar.
3. Add one spoonful of the salt to the cup and stir.
4. Find and record the temperature after adding the salt.
5. Observe for three minutes and record any observations.
6. Find and record the ending temperature.
7. Clean your equipment and area.

Reactions

Connecting Learning

1. What evidence do you have as to whether a chemical reaction took place at each station?

2. How are the temperature changes you observed in these investigations different than if you put something on the stove or in a refrigerator?

3. If you pour hydrogen peroxide on a cut and a white substance appears, what do you know has taken place?

4. Why was it important to record what you observed at each station?

5. How did the thermometer help determine if a chemical reaction had taken place?

6. Did some of the reactions have more than one thing happening that would tell you a chemical reaction has taken place? Explain.

7. What are you wondering now?

Mystery Reactions

Topic
Physical and Chemical Properties

Key Question
How can you identify a substance based on its physical and chemical properties?

Learning Goals
Students will:
1. make observations of eight white substances;
2. test the eight white substances for their reaction to vinegar, water, and iodine;
3, determine the pH for the eight white substances; and
4. predict the components of a mystery substance based on observations and tests.

Guiding Documents
Project 2061 Benchmarks
- *Scientific investigations may take many different forms, including observing what things are like or what is happening somewhere, collecting specimens for analysis, and doing experiments. Investigations can focus on physical, biological, and social questions.*
- *Heating and cooling cause changes in the properties of materials. Many kinds of changes occur faster under hotter conditions.*

NRC Standards
- *Use appropriate tools and techniques to gather, analyze, and interpret data.*
- *Think critically and logically to make the relationships between evidence and explanations.*
- *A substance has characteristic properties, such as density, a boiling point, and solubility, all of which are independent of the amount of the sample. A mixture of substances often can be separated into the original substances using one or more of the characteristic properties.*

Science
Physical science
    physical and chemical properties

Integrated Processes
Observing
Comparing and contrasting
Communicating

Predicting
Collecting and recording data
Interpreting data
Inferring

Materials
For each student group:
    eight empty film canisters (see *Management 1*)
    eight mystery substances (see *Management 2*)
    testing card (see *Management 3*)
    Mystery Substance sheet

For the class:
    station cards (see *Management 4*)

Background Information
    Scientists use the physical and chemical properties of materials to identify them. The physical properties of a substance are based on what the substance looks like such as its color or shape. This experience has the students examine the color of the white substances. The students should be able to see variation in the whiteness of the substances. Crystal shape is another physical characteristic that some of the white substances display.

    Chemical properties describe how a substance reacts chemically with another substance. This experience will examine how the mystery substances react with iodine, water, and vinegar. The final test the students will conduct is to use pH paper as an indicator of the pH values of the white substances. An indicator is a substance that changes color as an indication that two or more substances have reacted. The students will also approximate the pH level of the substance by reading a simplified pH chart. Chemists use pH to indicate the concentration of hydrogen ions in a solution. The letters pH stand for *potential of hydrogen*. Chemists use indicators to test whether a substance is an acid or a base. Indicators work by turning a distinctive color in the presence of an acid or a base.

Management
1. Obtain empty film canisters from a photo-developing store. Most stores will give these to you for no charge. Each group will need eight film canisters. Number (1–8) the canisters and lids with a permanent marking pen. If the canisters are black, use masking tape as labels on which to write the numbers.

2. Prepare a set of substances for each group of students. Place the following eight white substances into eight different containers:
Canister 1: salt; Canister 2: sugar; Canister 3: flour; Canister 4: baking soda; Canister 5: powdered lime; Canister 6: plaster of Paris; Canister 7: baking powder; and Canister 8: cornstarch. You should fill each container half full with the white substances. A small plastic basket or container will make it easier to carry the substances from station to station.

3. Laminate a *Testing Card* for each group. Make sure students know to only place the amount of substances needed to fill the circles on the *Testing Card*.

4. Prepare the six station cards. You will need to place the following equipment at different stations:
Station 1
    container of water
    eyedropper

Station 2
    container of iodine solution
    eyedropper

Station 3
    hand lenses or microscopes

Station 4
    pH paper
    container of water
    eyedropper
    pH scale

Station 5
    wooden clothespin
    aluminum foil
    heat source

Station 6
    container of vinegar
    eyedropper

4. The iodine solution can be prepared by placing a teaspoon of iodine into a cup of water.

5. Pet supply stores often carry the pH paper. It is used to test the salt water in fish tanks. Copy one pH scale for use at *Station 4*.

6. Powdered lime, also known as pickling lime, is available in grocery stores. Look for it where canning supplies are sold.

7. Calcium chloride is used to de-ice sidewalks during the winter. Most hardware/home supply stores carry it.

8. Remind the students not to taste or place any of the materials in their mouths.

9. For *Part Two*, you will need to mix two or three of the powders together. Do not tell students which powders have been mixed.

Procedure
Part One
1. Ask the *Key Question* and state the *Learning Goals*.
2. Distribute the eight containers of substances. Tell the students to remove the lids and look inside the containers. Remind the students to not mix the lids and containers.
3. Direct a discussion on the white substances that are in the containers. Have the students tell how the substances are the same and how they are different. Direct the students to first make color observations of the white substances. Tell them they will be conducting tests as well as making and recording observations so that they will learn some of the properties of each of the white substances.
4. Point out the location of the six station cards. Tell the students the order in which they will rotate through each station.
5. Distribute the *Testing Cards*, one to each group. Tell the students that they will fill the circles on the *Testing Card*, matching the substance from the canisters with the number by the circle. Inform them that they will need to wipe off their card with a damp paper towel after they have completed each station.
6. Monitor the progress of the students. Tell the students you will give a two-minute warning before they will need to rotate to the next station. Allow approximately five to ten minutes of time for each station rotation.
7. Direct a discussion on what the students observed and learned about each of the white substances.

Part Two
1. Ask the *Key Question* and state the *Learning Goals*.
2. Tell the students that they will now need to conduct the tests on a mystery substance that is a combination of at least two of the substances but not more than three. The students' task will be to determine what white substances—identified by the canister number—are in the mystery combination.
3. Direct the students through the rotation of the six stations for a second time. Point out that they will need to make very careful observations so that they will be a able to determine which of the eight substances are in the mystery sample.
4. Point out that how a substance doesn't react may be just as important as how it does react. For

example if it doesn't bubble when tested with the vinegar, they should know that lime and baking soda are not present.

Connecting Learning

1. Why was it necessary to conduct multiple tests for the white substances?
2. What did you learn about the color white?
3. For some substances, were some tests more important than others? Explain your thinking.
4. How did your group decide on what was in the mystery sample?
5. What do you think the difference is between physical properties and chemical properties?
6. What are you wondering now?

Mystery Reactions

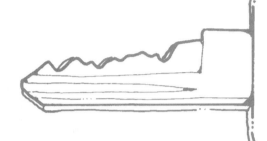

•Key Question•

How can you identify a substance based on its physical and chemical properties?

Learning Goals

Students will:

1. make observations of eight white substances;

2. test the eight white substances for their reaction to vinegar, water, and iodine;

3. determine the pH for the eight white substances; and

4. predict the components of a mystery substance based on observations and tests.

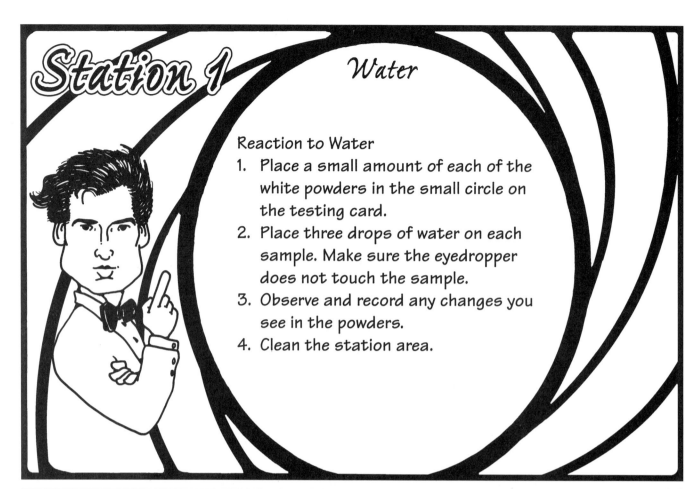

Station 1

Water

Reaction to Water
1. Place a small amount of each of the white powders in the small circle on the testing card.
2. Place three drops of water on each sample. Make sure the eyedropper does not touch the sample.
3. Observe and record any changes you see in the powders.
4. Clean the station area.

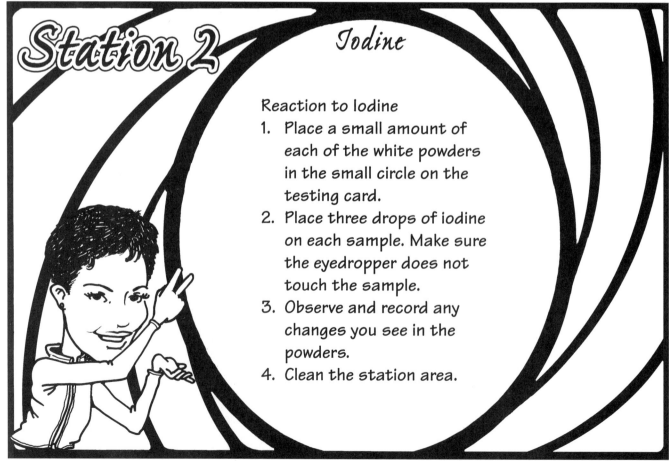

Station 2

Iodine

Reaction to Iodine
1. Place a small amount of each of the white powders in the small circle on the testing card.
2. Place three drops of iodine on each sample. Make sure the eyedropper does not touch the sample.
3. Observe and record any changes you see in the powders.
4. Clean the station area.

Station 3

Hand Lens or Microscope

Enhanced Observations
1. Place a small amount of each of the white powders in the small circle on the testing card.
2. Use the microscope or hand lens to make careful observations of the powders.
3. Clean the station area.

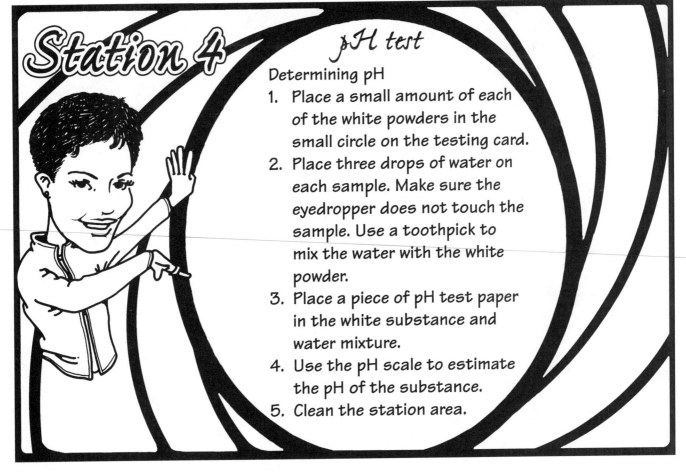

Station 4

pH test

Determining pH
1. Place a small amount of each of the white powders in the small circle on the testing card.
2. Place three drops of water on each sample. Make sure the eyedropper does not touch the sample. Use a toothpick to mix the water with the white powder.
3. Place a piece of pH test paper in the white substance and water mixture.
4. Use the pH scale to estimate the pH of the substance.
5. Clean the station area.

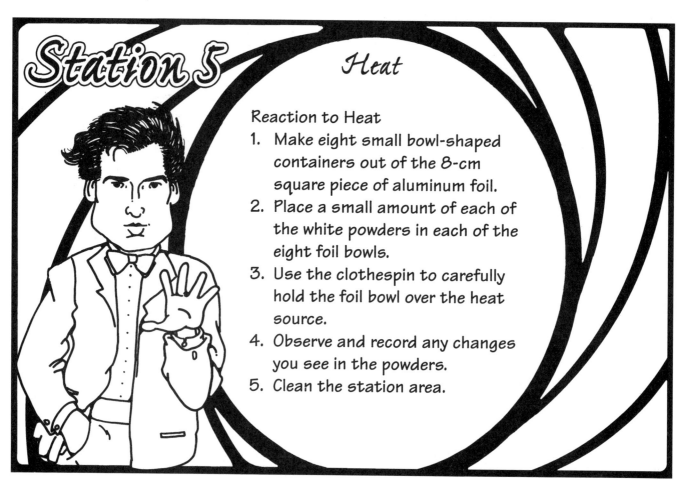

Station 5 — *Heat*

Reaction to Heat

1. Make eight small bowl-shaped containers out of the 8-cm square piece of aluminum foil.
2. Place a small amount of each of the white powders in each of the eight foil bowls.
3. Use the clothespin to carefully hold the foil bowl over the heat source.
4. Observe and record any changes you see in the powders.
5. Clean the station area.

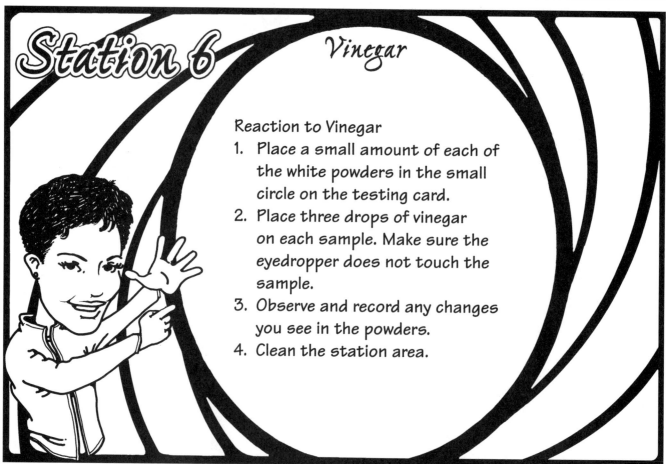

Station 6 — *Vinegar*

Reaction to Vinegar

1. Place a small amount of each of the white powders in the small circle on the testing card.
2. Place three drops of vinegar on each sample. Make sure the eyedropper does not touch the sample.
3. Observe and record any changes you see in the powders.
4. Clean the station area.

Mystery Reactions

Write your observations for each substance.

	Water	Iodine	Hand lens or microscope
1			
2			
3			
4			
5			
6			
7			
8			

Substance

Mystery Reactions

Write your observations for each substance.

	pH test	Heat	Vinegar	Color Analysis (How white is the substance?)
1				
2				
3				
4				
5				
6				
7				
8				

Substance

Mystery Reactions

Mystery Substance

These are the powders
I think are in the mystery substance:

pH Scale for Station 4

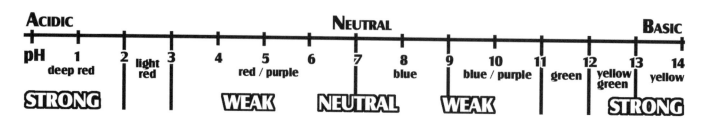

ACIDIC							NEUTRAL							BASIC

pH 1 deep red   2 light red   3   4   5 red / purple   6   7   8 blue   9   10 blue / purple   11   12 green   13 yellow green   14 yellow

STRONG   WEAK   NEUTRAL   WEAK   STRONG

Connecting Learning

1. Why was it necessary to conduct multiple tests for the white substances?

2. What did you learn about the color white?

3. For some substances, were some tests more important than others? Explain your thinking.

4. How did your group decide on what was in the mystery sample?

5. What do you think the difference is between physical properties and chemical properties?

6. What are you wondering now?

Topic
Periodic Table of the Elements

Key Question
What can we learn about organization that will apply to the periodic table of the elements?

Learning Goals
Students will:
1. sort and classify a set of eggs,
2. use a vertical and horizontal axis for classification, and
3. identify how to read parts of a periodic table of the elements.

Guiding Documents
Project 2061 Benchmarks
- *Organize information in simple tables and graphs and identify relationships they reveal.*
- *Read simple tables and graphs produced by others and describe in words what they show.*

NRC Standard
- *An element is composed of a single type of atoms. When elements are listed in order according to the number of protons (called the atomic number), repeating patterns of physical and chemical properties identify families of elements with similar properties. This "Periodic Table" is a consequence of the repeating pattern of the outermost electrons and their permitted energies.*

Science
Physical science
   periodic table

Integrated Processes
Observing
Relating
Communicating
Predicting
Collecting and organizing data

Materials
For each student group:
   a set of periodic eggs (see *Management 1*)
   an egg box sheet
   scissors

For each student:
   a periodic table
   periodic table information

Background Information
The organization within the periodic table of the elements is based on properties. The elements are grouped by families as well as by periods. The focus of this experience is to help students explore how the periodic table is organized along a vertical and horizontal axis. The periodic table that they will use is a simplified version that includes the atomic number as well as the atomic symbol. More complex information such as atomic mass and valances are not included. The focus is not to have the students memorize the table, but to help them to learn to read information from the table.

Management
1. Prepare a cardstock set of the periodic eggs and one egg box sheet for each group. Save the half page sheet of eggs for *Part Two*. These are the mystery eggs.
2. Prepare a copy of the periodic table and the periodic table information for each student.

Procedure
Part One
1. Have students get into groups. Ask the *Key Question* and state the *Learning Goals*.
2. Distribute the pictures of the periodic eggs and have students cut them out. Ask a student in each group to sort the eggs based on an observable property. Have the other students predict the criteria used to group the eggs. Have each student in the group sort the eggs at least once.
3. Discuss ways that the students grouped the pictures of the eggs. Point out that there are multiple ways the eggs can be sorted.
4. Draw the students' attention to the bands and the jewels on the eggs. Tell them that the bands are the horizontal rows with jewels. Ask the students to find the egg that has the least number of bands and the least number of jewels.
5. Distribute the egg box page and ask them to place the egg with the least number of bands and jewels in the first container of the egg box. Tell them that they must now sort the remaining eggs based on the rules of the box. The horizontal rule is the number of jewels arranged from fewest to most

going from left to right. Each row needs to contain the same type of jewel. The vertical rule is the number of bands, fewest to most going from top to bottom.

6. Check the students' organization of the eggs and discuss how they went about organizing the eggs.

7. Distribute a copy of the periodic table. Point out the periodic table is also organized along a vertical and horizontal axis based on the properties of the elements.

8. Hand out the periodic table information page for students to read. Focus on learning to read and interpret the table, not memorizing the elements.

Part Two

1. Tell the students that Dimitri Mendeleev, a Russian chemist, was the first to publish the classification of the elements that gave us the format we use today. Mendeleev organized his chart by patterns in the elements that he knew at that time. He predicted that some elements had not yet been discovered because there were spaces in his organizational chart that were not filled.

2. Tell the students that you have some mystery eggs that need to be classified. Distribute the half sheets of mystery eggs and encourage students to place them according to the patterns they have established in their charts.

3. Invite students to use the blank egg shapes to illustrate eggs that are missing.

4. Discuss the procedure and the results.

Connecting Learning

1. How did your group go about sorting the eggs?
2. How is the box of periodic eggs like a periodic table of the elements? How is it different?
3. What did you have to think about when placing the mystery eggs?
4. Did everyone position the mystery eggs in the same place? Explain.
5. Why do you think scientists organized the table of the elements this way?
6. What other things are organized into tables?
7. Why are tables useful in organizing data?
8. What are you wondering now?

Solutions

The diagram below shows how students should organize their eggs in the box. It also shows where the additional eggs from *Part Two* should be placed outside the box.

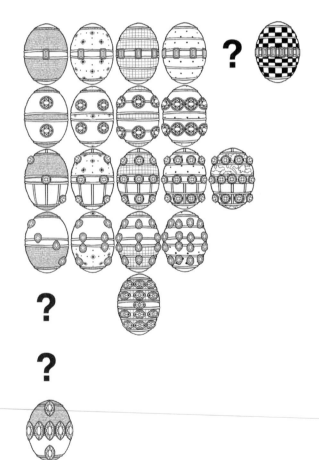

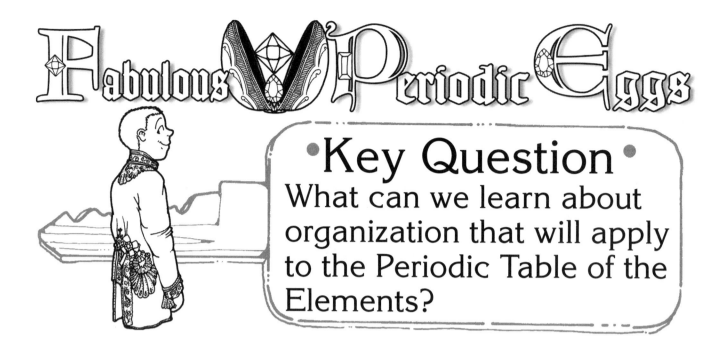

Learning Goals

STUDENTS WILL:

1. sort and classify a set of eggs,

2. use a vertical and horizontal axis for classification, and

3. identify how to read parts of a periodic table of the elements.

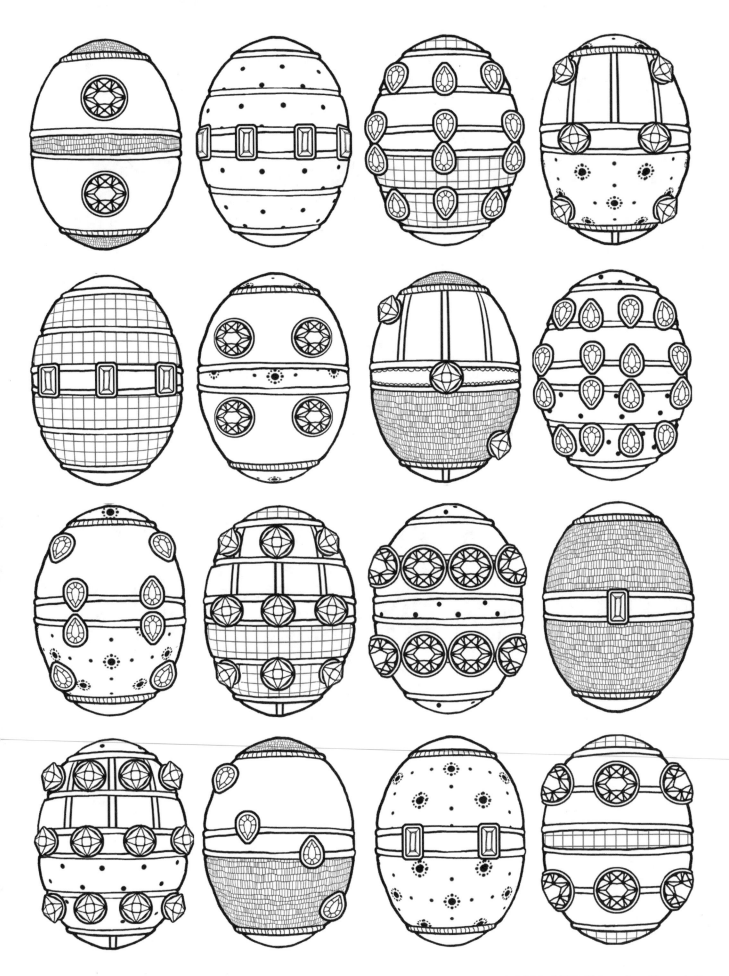

Periodic Table of the Elements

Group

1 **H** Hydrogen 1.00	Atomic Number Symbol Name Atomic Mass

○ Synthetically Prepared

● **Solid**

● **Liquid**

○ **Gas**

1 **H** Hydrogen 1.00																	2 **He** Helium 4.00
3 **Li** Lithium 6.94	4 **Be** Beryllium 9.01											5 **B** Boron 10.81	6 **C** Carbon 12.01	7 **N** Nitrogen 14.00	8 **O** Oxygen 15.99	9 **F** Fluorine 18.99	10 **Ne** Neon 20.18
11 **Na** Sodium 22.99	12 **Mg** Magnesium 24.30											13 **Al** Aluminum 26.98	14 **Si** Silicon 28.08	15 **P** Phosphorus 30.97	16 **S** Sulfur 32.06	17 **Cl** Chlorine 35.45	18 **Ar** Argon 39.94
19 **K** Potassium 39.09	20 **Ca** Calcium 40.07	21 **Sc** Scandium 44.95	22 **Ti** Titanium 47.86	23 **V** Vanadium 50.94	24 **Cr** Chromium 51.99	25 **Mn** Manganese 54.93	26 **Fe** Iron 55.84	27 **Co** Cobalt 58.93	28 **Ni** Nickel 58.69	29 **Cu** Copper 63.54	30 **Zn** Zinc 65.39	31 **Ga** Gallium 69.72	32 **Ge** Germanium 72.61	33 **As** Arsenic 74.92	34 **Se** Selenium 78.96	35 **Br** Bromine 79.90	36 **Kr** Krypton 83.80
37 **Rb** Rubidium 85.46	38 **Sr** Strontium 87.62	39 **Y** Yttrium 88.90	40 **Zr** Zirconium 91.22	41 **Nb** Niobium 92.90	42 **Mo** Molybdenum 95.94	43 **Tc** Technetium (98)	44 **Ru** Ruthenium 101.07	45 **Rh** Rhodium 102.91	46 **Pd** Palladium 106.42	47 **Ag** Silver 107.87	48 **Cd** Cadmium 112.41	49 **In** Indium 114.82	50 **Sn** Tin 118.71	51 **Sb** Antimony 121.76	52 **Te** Tellurium 127.60	53 **I** Iodine 126.90	54 **Xe** Xenon 131.29
55 **Cs** Cesium 132.91	56 **Ba** Barium 137.33	57 **La** Lanthanum 138.91	72 **Hf** Hafnium 178.49	73 **Ta** Tantalum 180.95	74 **W** Tungsten 183.84	75 **Re** Rhenium 186.21	76 **Os** Osmium 190.23	77 **Ir** Iridium 192.22	78 **Pt** Platinum 195.08	79 **Au** Gold 196.97	80 **Hg** Mercury 200.59	81 **Tl** Thallium 204.38	82 **Pb** Lead 207.20	83 **Bi** Bismuth 208.98	84 **Po** Polonium (209)	85 **At** Astatine (210)	86 **Rn** Radon (222)
87 **Fr** Francium (223)	88 **Ra** Radium (226)	89 **Ac** Actinium (227)	104 **Rf** Rutherfordium (261)	105 **Db** Dubnium (262)	106 **Sg** Seaborgium (263)	107 **Bh** Bohrium (264)	108 **Hs** Hassium (265)	109 **Mt** Meitnerium (268)	110 **Uun** Ununnilium (281)	111 **Uuu** Unununium (272)	112 **Uub** Ununbium (285)	113 **Uut** Ununtrium (282)	114 **Uuq** Ununquadium (289)	115 **Uup** Ununpentium (289)	116 **Uuh** Ununhexium (289)	117 **Uus** Ununseptium (293)	118 **Uuo** Ununoctium (293)

58 **Ce** Cerium 140.12	59 **Pr** Praseodymium 140.91	60 **Nd** Neodymium 144.24	61 **Pm** Promethium (145)	62 **Sm** Samarium 150.36	63 **Eu** Europium 151.96	64 **Gd** Gadolinium 157.25	65 **Tb** Terbium 158.93	66 **Dy** Dysprosium 162.50	67 **Ho** Holmium 164.93	68 **Er** Erbium 167.26	69 **Tm** Thulium 168.93	70 **Yb** Ytterbium 173.04	71 **Lu** Lutetium 174.97
90 **Th** Thorium 232.04	91 **Pa** Protactinium 231.04	92 **U** Uranium 238.03	93 **Np** Neptunium (237)	94 **Pu** Plutonium (244)	95 **Am** Americium (243)	96 **Cm** Curium (247)	97 **Bk** Berkelium (247)	98 **Cf** Californium (251)	99 **Es** Einsteinium (252)	100 **Fm** Fermium (257)	101 **Md** Mendelevium (258)	102 **No** Nobelium (259)	103 **Lr** Lawrencium (262)

Period

Periodic Table of the Elements

During the time of Aristotle, people thought that everything was made up of a combination of air, fire, water, and earth. Today we know that there are a certain number of elements that make up all matter on Earth. These elements are composed of tiny particles called atoms. Every atom is, in turn, composed of a certain number of protons, neutrons, and electrons. Once it was thought that protons, neutrons, and electrons were the smallest components of matter, but recent discoveries show that even they can be broken into smaller particles.

In the late 1860s, a Russian chemist named Dmitri Mendeleev began to work on organizing the elements based on patterns. He published his first periodic table of the elements in 1869. This work is the basis for our current periodic table of the elements in which elements are grouped by properties. Each element is represented by a symbol of one or two letters that stands for its name. Chemists use these symbols to write formulas for compounds. The elements are listed in rows horizontally in order of their atomic number and in columns vertically in families, or groups. These families share similar chemical properties.

Each element has its own square on the chart. Most periodic tables are organized in similar manners. The large number at the top of the square is called the Atomic Number. It stands for the number of protons in the nucleus. It also tells how many electrons are in the element. Equal numbers of protons and electrons mean that the charge of the atom is neutral or balanced. The number at the bottom of the square is the atomic weight. This number is used to determine the number of neutrons that are in an element. Oxygen's atomic weight is 15.99. To determine how many neutrons are in this element, you round 15.99 to 16 and subtract the atomic number (8) from it: $16 - 8 = 8$. This tells you that oxygen has 8 neutrons.

26
Fe
Iron
55.84

1
H
Hydrogen
1.00

27
Co
Cobalt
58.93

79
Au
Gold
196.97

28
Ni
Nickel
58.69

30
Zn
Zinc
65.39

11
Na
Sodium
22.99

20
Ca
Calcium
40.07

Connecting Learning

1. How did your group go about sorting the eggs?

2. How is the box of periodic eggs like a periodic table of the elements? How is it different?

3. What did you have to think about when placing the mystery eggs?

4. Did everyone position the mystery eggs in the same place? Explain.

5. Why do you think scientists organized the table of the elements this way?

6. What other things are organized into tables?

7. Why are tables useful in organizing data?

8. What are you wondering now?

It's Elemental, My Dear

Topic
Periodic Table of the Elements

Key Questions
1. What are the most common elements in the Earth's crust?
2. What do chemical formulas tell about the elements in molecules?

Learning Goals
Students will:
1. read and interpret tables,
2. construct a circle graph, and
3. read a periodic table to identify elements found in the molecules of selected minerals.

Guiding Documents
Project 2061 Benchmarks
- *Materials may be composed of parts that are too small to be seen without magnification.*
- *When a new material is made by combining two or more materials, it has properties that are different from the original materials. For that reason, a lot of different materials can be made from a small number of basic kinds of materials.*

NRC Standard
- *Chemical elements do not break down during normal laboratory reactions involving such treatments as heating, exposure to electric current, or reaction with acids. There are more than 100 known elements that combine in a multitude of ways to produce compounds, which account for the living and nonliving substances that we encounter.*

NCTM Standards 2000*
- *Recognize and generate equivalent forms of commonly used fractions, decimals, and percents*
- *Collect data using observations, surveys, and experiments*

Math
Data analysis
circle graphs

Science
Physical science
chemistry
elements

Integrated Processes
Observing
Comparing and contrasting
Communicating
Organizing data
Interpreting data
Analyzing data

Materials
Student pages
Colored pencils or crayons
Protractors

Background Information
In this activity, students are encouraged to read in the content area for a purpose. The activity is divided into two parts. The first portion has students construct a circle graph that shows the percentages of eight of the most common elements found in the Earth's crust. The circle helps students see part-to-whole relationships. The second part of the activity directs students to use a periodic table to identify elements found in selected molecules of minerals. Students should understand that some minerals, as well as other substances, are composed of single elements such as iron (Fe); however, most minerals, as well as most other substances, are composed of different combinations of elements. H_2O is the chemical symbol for a single molecule of water. It is composed of two elements, hydrogen and oxygen. The formula also tells us that there are two hydrogen atoms and one oxygen atom in one molecule of water.

Management
1. It is assumed that students know how to use a protractor to divide a circle into sections.

Procedure

1. Ask the *Key Question* and state the *Learning Goals*.
2. Distribute the student page that has *Table One* on it. Direct the students to read the information. Discuss with the students the information from the passage:
   - What are the most common elements in the Earth's crust?
   - What do you think the word *crystalline* means?
   - What are some other examples of inorganic substances?
   - What do you think the value of gold would be if it made up 50% of the Earth's crust?
   - Why do you think the name of this activity is *It's Elemental, My Dear?*
3. Distribute the circle graph and protractor. Tell the students to create a circle graph from the data in *Table One*. (The data only account for 99% of the crust. Help the students to understand that the other 1% is all the other elements.)
4. Distribute the remaining student pages. Tell students to read the passage about minerals. Ask the students to use the periodic table to explain where the Au and Ag come from. [In the periodic table, abbreviations are used to represent elements. Au is the symbol for gold, and Ag is the symbol for silver.]
5. Direct the students to select eight minerals from *Table Two* to analyze. Have them look at the example and ask them if they can explain how the number of atoms is determined. [The symbol tells the type of atom and the subscript number tells the amount. In the example, one atom of calcium and one atom of magnesium are listed because when there is no subscript, only one atom is represented. There are two atoms of carbon and six atoms of oxygen. You have six atoms of oxygen because 3 subscript times 2 subscript equals six ($3 \cdot 2 = 6$). There are 2 atoms of carbon for the same reason ($1 \cdot 2 = 2$).]
6. Assist students as necessary as they analyze the minerals they selected.

Connecting Learning

1. How did the circle graph help you better see the data in *Table One*?
2. How did the *Periodic Table* help you in completing the chart?
3. What elements are in H_2O? [hydrogen and oxygen] How do you know? How many atoms would there be in H_2O_2? [four, two atoms of hydrogen and two atoms of oxygen] How do you know?
4. The mineral hornblende is a very heavy mineral. Based on what you learned in this activity, explain why this may be so? [Each molecule of hornblende has 59 atoms in it.]
5. What information in *Table Two* would lead you to believe that oxygen is a very common element in the Earth's crust? [Oxygen is a common element in many minerals.]
6. What are you wondering now?

* Reprinted with permission from *Principles and Standards for School Mathematics*, 2000 by the National Council of Teachers of Mathematics. All right reserved.

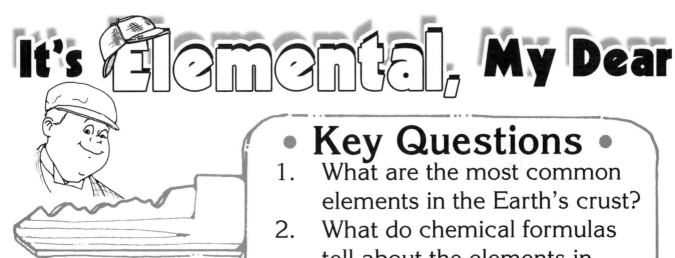

It's Elemental, My Dear

Learning Goals

STUDENTS WILL:

1. read and interpret tables,
2. construct a circle graph,
3. read a periodic table to identify elements found in the molecules of selected minerals.

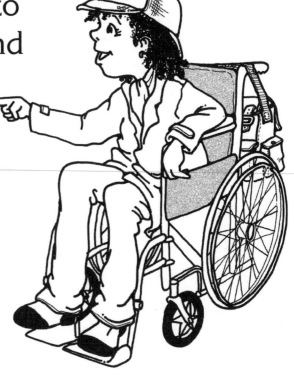

It's Elemental, My Dear

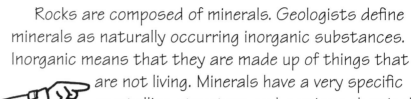

Rocks are composed of minerals. Geologists define minerals as naturally occurring inorganic substances. Inorganic means that they are made up of things that are not living. Minerals have a very specific crystalline structure and a unique chemical composition. The minerals found in the Earth's rocks are produced by different arrangements of chemical elements. A list of the eight most common elements that make up the minerals found in the Earth's crust are described in *Table One*.

	Element	Chemical Symbol	Percentage of the Earth's Crust
Table One	Oxygen	O	47%
	Silicon	Si	28%
	Aluminum	Al	8%
	Iron	Fe	5%
	Calcium	Ca	4%
	Sodium	Na	3%
	Potassium	K	2%
	Magnesium	Mg	2%

Construct a circle graph that shows the percentage of each of the eight most common minerals. Create a key for the circle graph.

It's Elemental, My Dear

Construct a circle graph and key of Earth's eight most common minerals.

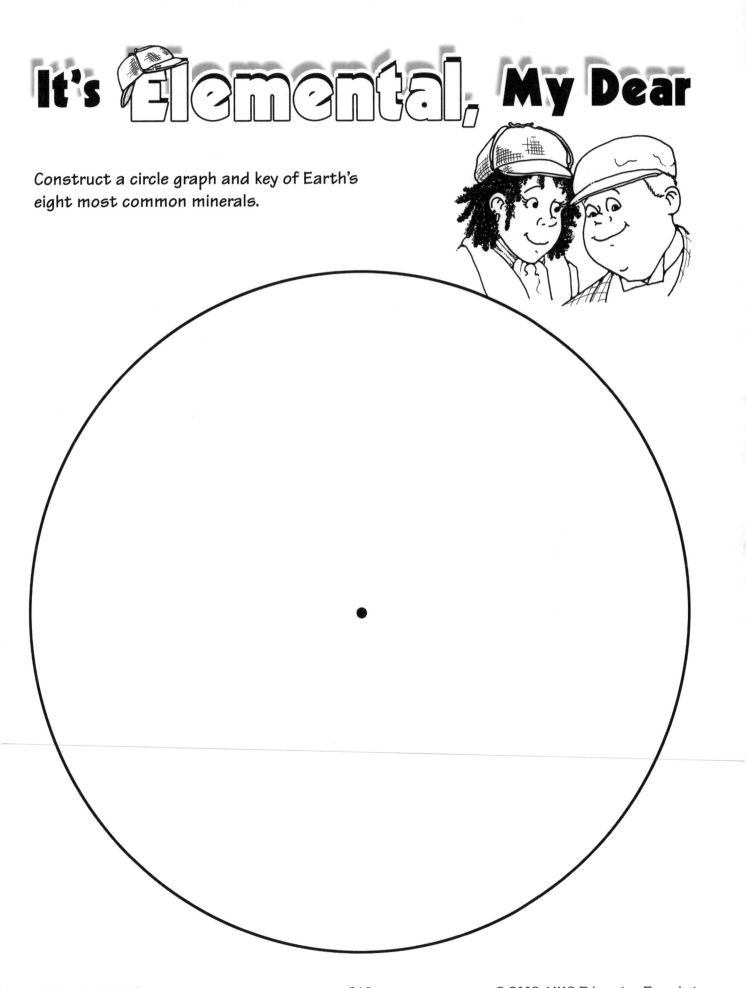

It's Elemental, My Dear

Over 2000 minerals have been identified by geologists. They classify minerals into groups based on how they are similar. *Table Two* lists some of these groups. The *Elements* group includes over 100 known minerals. Many of the minerals in this group are composed of only one element. Geologists sometimes subdivide this group into metal and nonmetal categories. Gold, silver, and copper are examples of metals. The elements sulfur and carbon are nonmetallic minerals. Most minerals are combinations of two or more elements. *Table Two* describes the chemical composition of some common minerals. The chemical composition lists the elements that make up the minerals. For example, the chemical composition for the mineral quartz is SiO_2. That means each quartz molecule is composed of one atom of silicon and two atoms of oxygen. Each molecule of quartz has three atoms in it.

Select eight minerals from *Table Two*. Select at least one from each group. Use a periodic table to identify the elements found in one molecule of that mineral. You will also need to identify the number of atoms found each of the elements in that mineral. Record this information on your student sheet. An example has been done for you.

Table Two

Group	Mineral	Chemical Composition
Elements	Gold	Au
	Silver	Ag
	Copper	Cu
	Carbon	C
	Sulfur	S
Sulfides	Galena	PbS
	Pyrite	FeS_2
Halides	Fluorite	CaF_2
	Halite	NaCl
Oxides	Corundum	Al_2O_3
	Hematite	Fe_2O_3
Carbonites	Calcite	$CaCO_3$
	Dolomite	$CaMg(CO_3)_2$
Sulfates	Gypsum	$CaSO_4(H_2O)$
Silicates	Beryl	$Be_3Al_2(SiO_3)_6$
	Biotite	$K(Mg,Fe)_3(Si_3O_{10})(OH)_2$
	Hornblende	$Ca_2(Mg,Fe,Al)_5(AlSi)_8O_{22}(OH)_2$
	Muscovite	$KAl_3Si_3O_{10}(OH)_2$
	Olivine	$(Mg,Fe)_2SiO_4$
	Quartz	SiO_2

It's Elemental, My Dear

Mineral

Elements

Example	
Dolomite $CaMg(CO_3)_2$	1 atom of calcium, 1 atom of magnesium, 2 atoms of carbon, and six atoms of oxygen

Periodic Table of the Elements

Group

1																	**2**
H Hydrogen 1.00																	**He** Helium 4.00

Legend:
- Atomic Number
- Symbol
- Name
- Atomic Mass

1 **H** Hydrogen 1.00

- ○ Synthetically Prepared
- ● **Solid**
- ◗ **Liquid**
- ◯ **Gas**

Period																	
3 Li Lithium 6.94	**4 Be** Beryllium 9.01											**5 B** Boron 10.81	**6 C** Carbon 12.01	**7 N** Nitrogen 14.00	**8 O** Oxygen 15.99	**9 F** Fluorine 18.99	**10 Ne** Neon 20.18
11 Na Sodium 22.99	**12 Mg** Magnesium 24.30											**13 Al** Aluminum 26.98	**14 Si** Silicon 28.08	**15 P** Phosphorus 30.97	**16 S** Sulfur 32.06	**17 Cl** Chlorine 35.45	**18 Ar** Argon 39.94
19 K Potassium 39.09	**20 Ca** Calcium 40.07	**21 Sc** Scandium 44.95	**22 Ti** Titanium 47.86	**23 V** Vanadium 50.94	**24 Cr** Chromium 51.99	**25 Mn** Manganese 54.93	**26 Fe** Iron 55.84	**27 Co** Cobalt 58.93	**28 Ni** Nickel 58.69	**29 Cu** Copper 63.54	**30 Zn** Zinc 65.39	**31 Ga** Gallium 69.72	**32 Ge** Germanium 72.61	**33 As** Arsenic 74.92	**34 Se** Selenium 78.96	**35 Br** Bromine 79.90	**36 Kr** Krypton 83.80
37 Rb Rubidium 85.46	**38 Sr** Strontium 87.62	**39 Y** Yttrium 88.90	**40 Zr** Zirconium 91.22	**41 Nb** Niobium 92.90	**42 Mo** Molybdenum 95.94	**43 Tc** Technetium (98)	**44 Ru** Ruthenium 101.07	**45 Rh** Rhodium 102.91	**46 Pd** Palladium 106.42	**47 Ag** Silver 107.87	**48 Cd** Cadmium 112.41	**49 In** Indium 114.82	**50 Sn** Tin 118.71	**51 Sb** Antimony 121.76	**52 Te** Tellurium 127.60	**53 I** Iodine 126.90	**54 Xe** Xenon 131.29
55 Cs Cesium 132.91	**56 Ba** Barium 137.33	**57 La** Lanthanum 138.91	**72 Hf** Hafnium 178.49	**73 Ta** Tantalum 180.95	**74 W** Tungsten 183.84	**75 Re** Rhenium 186.21	**76 Os** Osmium 190.23	**77 Ir** Iridium 192.22	**78 Pt** Platinum 195.08	**79 Au** Gold 196.97	**80 Hg** Mercury 200.59	**81 Tl** Thallium 204.38	**82 Pb** Lead 207.20	**83 Bi** Bismuth 208.98	**84 Po** Polonium (209)	**85 At** Astatine (210)	**86 Rn** Radon (222)
87 Fr Francium (223)	**88 Ra** Radium (226)	**89 Ac** Actinium (227)	**104 Rf** Rutherfordium (261)	**105 Db** Dubnium (262)	**106 Sg** Seaborgium (263)	**107 Bh** Bohrium (264)	**108 Hs** Hassium (265)	**109 Mt** Meitnerium (268)	**110 Uun** Ununnilium (281)	**111 Uuu** Unununium (272)	**112 Uub** Ununbium (285)	**113 Uut** Ununtrium (282)	**114 Uuq** Ununquadium (289)	**115 Uup** Ununpentium (289)	**116 Uuh** Ununhexium (289)	**117 Uus** Ununseptium (293)	**118 Uuo** Ununoctium (293)

58 Ce Cerium 140.12	**59 Pr** Praseodymium 140.91	**60 Nd** Neodymium 144.24	**61 Pm** Promethium (145)	**62 Sm** Samarium 150.36	**63 Eu** Europium 151.96	**64 Gd** Gadolinium 157.25	**65 Tb** Terbium 158.93	**66 Dy** Dysprosium 162.50	**67 Ho** Holmium 164.93	**68 Er** Erbium 167.26	**69 Tm** Thulium 168.93	**70 Yb** Ytterbium 173.04	**71 Lu** Lutetium 174.97
90 Th Thorium 232.04	**91 Pa** Protactinium 231.04	**92 U** Uranium 238.03	**93 Np** Neptunium (237)	**94 Pu** Plutonium (244)	**95 Am** Americium (243)	**96 Cm** Curium (247)	**97 Bk** Berkelium (247)	**98 Cf** Californium (251)	**99 Es** Einsteinium (252)	**100 Fm** Fermium (257)	**101 Md** Mendelevium (258)	**102 No** Nobelium (259)	**103 Lr** Lawrencium (262)

Connecting Learning

1. How did the circle graph help you better see the data in *Table One?*

2. How did the *Periodic Table* help you in completing the chart?

3. What elements are in H_2O? How do you know? How many atoms would there be in H_2O_2? How do you know?

4. The mineral hornblend is a very heavy mineral. Based on what you learned in this activity, explain why this may be so.

5. What information in *Table Two* would lead you to believe that oxygen is a very common element in the Earth's crust?

6. What are you wondering now?

Black Boxes

Topic
Atomic and Molecular Structure

Key Question
How do scientists know what molecules and atoms look like?

Learning Goals
Students will:
1. construct an inference box,
2. infer the contents of boxes, and
3. relate how scientists develop models for materials that they cannot directly see.

Guiding Documents
Project 2061 Benchmark
- *All matter is made up of atoms, which are far too small to see directly through a microscope. The atoms of any element are alike but are different from atoms of other elements. Atoms may stick together in well-defined molecules or may be packed together in large arrays. Different arrangements of atoms into groups compose all substances.*

NRC Standards
- *Different kinds of questions suggest different kinds of scientific investigations. Some investigations involve observing and describing objects, organisms, or events; some involve collecting specimens; some involve experiments; some involve seeking more information; some involve discovery of new objects and phenomena; and some involve making models.*
- *Develop descriptions, explanations, predictions, and models using evidence.*

Science
Physical science
    chemistry
        atomic structure

Integrated Processes
Observing
Comparing and contrasting
Relating
Communicating
Predicting
Collecting and recording data
Inferring

Materials
For each student group:
    1 box that can be sealed (see *Management 1*)

For the class:
    1 prepared inference box (see *Management 2*)

Background Information
Science has shown us that some things such as atoms and molecules have parts that are too small to be directly seen. These parts have been identified based on evidence on how they interact with other things. For example scientists have been able to observe evidence of the paths taken by the electrons around its nucleus although the electrons themselves have not been seen. The focus of this activity is for students to understand how inferences are used to learn about things that cannot be directly seen.

Management
1. Small mailing boxes can be used to make the inference boxes.
2. Prepare a box with three or four marbles inside. Seal the box so that it cannot be opened.

Procedure
1. Ask the *Key Question* and state the *Learning Goals*.
2. Allow each group to observe the inference box you have prepared. Ask them what they think is inside the box.
3. Allow the class time to discuss what they think is in the box and explain what influenced their thinking.
4. Lead the students in a discussion on how scientists sometimes use evidence based on what they know or how something acts to infer what something looks like. Tell the students that just like scientists cannot see an atom, they will also not be allowed to see inside the box. They will need to identify what is in the box based on inferences.

5. Direct each student group to construct their own black box that other groups will observe and infer the contents. You may want to limit the number of items placed in the boxes.

6. Allow students time to construct group black boxes and time for each group to make inferences about the contents of each group's black box. Direct the students to write as well as draw what they think is in each group's black box.

7. Be sure to ask groups not to tell what they put in their boxes—even after others have stated their inferences.

Connecting Learning

1. What things did you think were in the boxes?
2. What influenced your thinking?
3. How did your group work out any differences of opinions of what was in an individual box?
4. How did this activity help you understand how scientists know what atoms and molecules look like?
5. When else do you use inferences?
6. What are you wondering now?

Black Boxes

Key Question

How do scientists know what molecules and atoms look like?

Learning Goals

STUDENTS WILL:

1. construct an inference box,
2. infer the contents of boxes, and
3. relate how scientists develop models for materials that they cannot directly see.

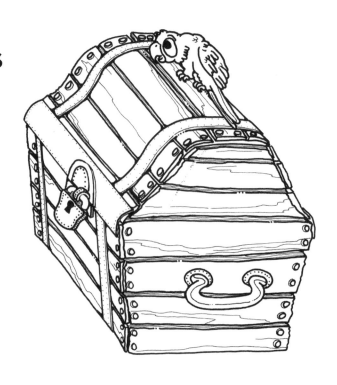

This is what I think is inside...

Black Boxes

Connecting Learning

1. What things did you think were in the boxes?

2. What influenced your thinking?

3. How did your group work out any differences of opinions of what was in an individual box?

4. How did this activity help you understand how scientists know what atoms and molecules look like?

5. When else do you use inferences?

6. What are you wondering now?

Topic
Modeling Atomic Structure

Key Question
What can we learning about atoms by studying different models of them?

Learning Goals
Students will:
1. examine different models of atoms,
2. identify strengths and weaknesses of each model, and
3. construct models to understand some atomic structures.

Guiding Documents
Project 2061 Benchmarks
- *Scientific ideas about elements were borrowed from some Greek philosophers of 2,000 years earlier, who believed that everything was made from our basic substances: air, earth, fire, and water. It was the combinations of these "elements" in different proportions that gave other substances their observable properties. The Greeks were wrong about those four, but now over 100 different elements have been identified, some rare and some plentiful, out of which everything is made. Because most elements tend to combine with others, few elements are found in their pure form.*
- *Different models can be used to represent the same thing. What kind of a model to use and how complex it should be depends on its purpose. The usefulness of a model may be limited if it is too simple or if it is needlessly complicated. Choosing a useful model is one of the instances in which intuition and creativity come into play in science, mathematics, and engineering.*

NRC Standard
- *Chemical elements do not break down during normal laboratory reactions involving such treatments as heating, exposure to electric current, or reaction with acids. There are more than 100 known elements that combine in a multitude of ways to produce compounds, which account for the living and nonliving substances that we encounter.*

Science
Physical science
  atomic structure

Integrated Processes
Observing
Comparing and contrasting
Communicating
Predicting
Inferring

Materials
For each student group:
  red, white, and blue dots (see *Management 1*)
  red, white, and blue beads, 8 mm
  3 pieces of pipe cleaner, 15 cm long
  3 pieces of clay in three different colors
  1 clear plastic craft ball (see *Management 2*)
  1 piece of clear plastic wrap, 10 cm square

Background Information
Every object we observe is composed of matter. Matter is defined by scientists as anything that has both mass and takes up space. All matter is composed of atoms. Scientists have discovered that all atoms are made up of smaller particles called protons, neutrons, and electrons. The protons and neutrons are contained in the nucleus, and the electrons orbit around the nucleus. Scientist use models to help them explain and explore things that are too large or too small to be directly observed.

Management
1. Use a hole punch to make the red, white, and blue dots to use on the first model.
2. Clear plastic craft balls can be purchased at stores that carry craft supplies. A craft ball with a diameter of 8 cm works well. You will need to select the type that snaps apart into two hemispheres.

Procedure
Part One
1. Ask the *Key Question* and state the *Learning Goals*.
2. Explain to the students that they will be building three different models of an atom. Each model will show some different characteristics of atoms.
3. Review with the students the following information about atoms. Point out that scientists have discovered that atoms have three parts: protons, neutrons, and electrons. The protons and neutrons are in the center or the nucleus of the atom and the electrons orbit the nucleus.

Direct the students' attention to the periodic table. Review how to determine the number of electrons, protons, and neutrons in an atom.

4. Distribute the activity sheets and construction paper dots. Direct the students in the construction of the first model, a model of a carbon atom. Tell them the blue dots represent the protons, the white dots represent the neutrons, and the red dots represent the electrons. Direct the students to glue 6 neutrons and 6 protons in the middle of the drawing. Tell them to equally space two electrons on the first energy ring and the others on the second ring and glue them down.

CARBON

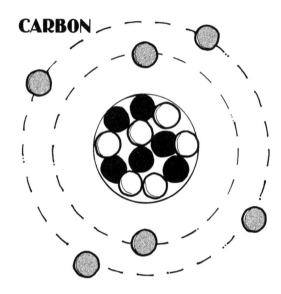

5. Point out to the students that they have just constructed a two-dimensional model of an atom. Next, they will construct a three-dimensional model.

Part Two

1. Ask the *Key Question* and state the *Learning Goals*. Distribute the beads and the pipe cleaners. Tell the students they will use red beads to represent the electrons, white beads for neutrons, and blue beads for protons. The pipe cleaners will be used to cluster the neutrons and protons in the nucleus as well as to display the electrons around the nucleus.

2. Tell the students they will be constructing a model for the element nitrogen. Tell them to use the periodic table to determine how many beads they will need for their model. Demonstrate how to string the beads that represent the neutrons and protons on the pipe cleaner and wrap it into a small circle.

3. Direct the students to use the second and third pipe cleaners to string the beads that represent the electrons. Tell the students to place the neutron and proton cluster on the table and to

create a ring around it with the second pipe cleaner with two electrons. Point out that they will need to space the beads that represent the electrons evenly around the circle. Then have them place the second ring of electrons on the outside. Tell them to glue this model on the activity sheet.

NITROGEN

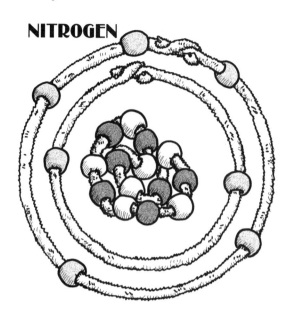

4. Have the students compare this model with the first model that they constructed. How are they the same and how are they different?

Part Three

1. Ask the *Key Question* and state the *Learning Goals*.

2. Review the two models that the students have just constructed. Explain that they will now construct a third model.

3. Show the students the clear plastic sphere. Tell them they will be using this and clay to build a third model of an atom.

4. Ask the students what parts of the atom they will need to show. [protons, neutrons, and electrons]

5. Distribute the three different colors of clay to the student groups. Direct them to locate the element boron on the periodic table. Have them tell how many electrons, neutrons, and protons they will need to show in the model.

6. Demonstrate how to roll the clay to create small balls of clay in each of the three colors.

7. Have the students assign the three parts of the atom to the different clay colors.

8. Show them how to place the clay spheres that represent the protons and neutrons into a cluster. Tell them to place the clay cluster on the clear plastic wrap and to suspend the plastic wrap between the two hemispheres. Tell them to connect the two hemispheres together so that

the clay cluster (the nucleus) is suspended in the middle of the clear craft ball.

9. Direct the students to place the clay balls that represent the electrons on the outside surface of the plastic craft sphere.

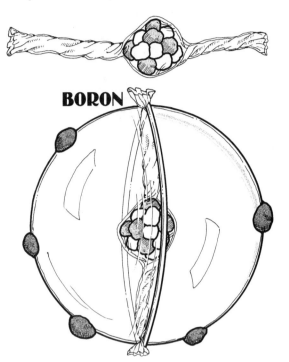

BORON

10. On the second activity sheet, have students compare and contrast the three different models. [The three different models have different strengths as well as weaknesses. The first model shows the parts of an atom in two-dimensional model. The parts cannot be moved to show the orbiting nature of the electrons around the nucleus. The second model show the parts of an atom in a more three-dimensional nature, but the orbiting electrons are contained in a single plane which is not how electrons are thought to orbit around the nucleus. The third model shows the structures of the atom in a three-dimensional nature, but the electrons are fixed on the plastic sphere.]

Connecting Learning

1. What are the three different parts of an atom?
2. How did the periodic table help you in this activity?
3. Why do scientists use models of atoms?
4. What do the models show? What do the models not show?
5. What will happen if we learn new things about the structure of atoms?
6. What are you wondering now?

Modeling ATOMS

•Key Question•

What can we learning about atoms by studying different models of them?

Learning Goals

STUDENTS WILL:

1. examine different models of atoms,

2. identify strengths and weaknesses of each model, and

3. construct models to understand some atomic structures.

ATOMIC *Make Up*

Every object in the universe is made up of matter. Matter is anything that has mass and takes up space. Matter is made up of atoms. Atoms contain even smaller parts called protons, neutrons, and electrons.

Scientists have developed a model of what the inside parts of atoms look like. In very simple terms an atom's parts are located in two places, the nucleus and in energy levels or shells that are around the nucleus. In the nucleus of every atom you can find one or more protons and neutrons. A proton is a positively charged particle and a neutron is a particle that has no charge. Electrons have a negative charge. Electrons orbit the nucleus in the shells or energy levels. An atom can have up to seven different energy levels. Each level is capable of holding different numbers of electrons.

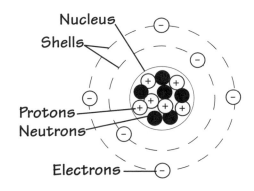

The attraction between the positively charged protons and negatively charged electrons is what holds the atom together. Atoms usually have the same number of electrons and protons—negative and positive charges—which causes them to have a neutral charge. The unlike charges cancel each other out.

Our understanding of the parts of atoms will probably change as we learn more about atoms. The big idea is that all atoms have an organized structure and we can use this information about their structure to build mental and physical models that will help us better understand what they look like.

Periodic Table of the Elements

Group

1																		2
1 **H** Hydrogen 1.00																		**2** **He** Helium 4.00

Key

1
H
Hydrogen
1.00

- Atomic Number — 1
- Symbol — H
- Name — Hydrogen
- Atomic Mass — 1.00

○ Synthetically Prepared
● Solid
● Liquid
○ Gas

| **3** Li Lithium 6.94 | **4** Be Beryllium 9.01 |
| **11** Na Sodium 22.99 | **12** Mg Magnesium 24.30 |

| **5** B Boron 10.81 | **6** C Carbon 12.01 | **7** N Nitrogen 14.00 | **8** O Oxygen 15.99 | **9** F Fluorine 18.99 | **10** Ne Neon 20.18 |
| **13** Al Aluminum 26.98 | **14** Si Silicon 28.08 | **15** P Phosphorus 30.97 | **16** S Sulfur 32.06 | **17** Cl Chlorine 35.45 | **18** Ar Argon 39.94 |

19 K Potassium 39.09	**20** Ca Calcium 40.07	**21** Sc Scandium 44.95	**22** Ti Titanium 47.86	**23** V Vanadium 50.94	**24** Cr Chromium 51.99	**25** Mn Manganese 54.93	**26** Fe Iron 55.84	**27** Co Cobalt 58.93	**28** Ni Nickel 58.69	**29** Cu Copper 63.54	**30** Zn Zinc 65.39	**31** Ga Gallium 69.72	**32** Ge Germanium 72.61	**33** As Arsenic 74.92	**34** Se Selenium 78.96	**35** Br Bromine 79.90	**36** Kr Krypton 83.80
37 Rb Rubidium 85.46	**38** Sr Strontium 87.62	**39** Y Yttrium 88.90	**40** Zr Zirconium 91.22	**41** Nb Niobium 92.90	**42** Mo Molybdenum 95.94	**43** Tc Technetium (98)	**44** Ru Ruthenium 101.07	**45** Rh Rhodium 102.91	**46** Pd Palladium 106.42	**47** Ag Silver 107.87	**48** Cd Cadmium 112.41	**49** In Indium 114.82	**50** Sn Tin 118.71	**51** Sb Antimony 121.76	**52** Te Tellurium 127.60	**53** I Iodine 126.90	**54** Xe Xenon 131.29
55 Cs Cesium 132.91	**56** Ba Barium 137.33	**57** La Lanthanum 138.91	**72** Hf Hafnium 178.49	**73** Ta Tantalum 180.95	**74** W Tungsten 183.84	**75** Re Rhenium 186.21	**76** Os Osmium 190.23	**77** Ir Iridium 192.22	**78** Pt Platinum 195.08	**79** Au Gold 196.97	**80** Hg Mercury 200.59	**81** Tl Thallium 204.38	**82** Pb Lead 207.20	**83** Bi Bismuth 208.98	**84** Po Polonium (209)	**85** At Astatine (210)	**86** Rn Radon (222)
87 Fr Francium (223)	**88** Ra Radium (226)	**89** Ac Actinium (227)	**104** Rf Rutherfordium (261)	**105** Db Dubnium (262)	**106** Sg Seaborgium (263)	**107** Bh Bohrium (264)	**108** Hs Hassium (265)	**109** Mt Meitnerium (268)	**110** Uun Ununnilium (281)	**111** Uuu Unununium (272)	**112** Uub Ununbium (285)	**113** Uut Ununtrium (282)	**114** Uuq Ununquadium (289)	**115** Uup Ununpentium	**116** Uuh Ununhexium (289)	**117** Uus Ununseptium	**118** Uuo Ununoctium (293)

| **58** Ce Cerium 140.12 | **59** Pr Praseodymium 140.91 | **60** Nd Neodymium 144.24 | **61** Pm Promethium (145) | **62** Sm Samarium 150.36 | **63** Eu Europium 151.96 | **64** Gd Gadolinium 157.25 | **65** Tb Terbium 158.93 | **66** Dy Dysprosium 162.50 | **67** Ho Holmium 164.93 | **68** Er Erbium 167.26 | **69** Tm Thulium 168.93 | **70** Yb Ytterbium 173.04 | **71** Lu Lutetium 174.97 |
| **90** Th Thorium 232.04 | **91** Pa Protactinium 231.04 | **92** U Uranium 238.03 | **93** Np Neptunium (237) | **94** Pu Plutonium (244) | **95** Am Americium (243) | **96** Cm Curium (247) | **97** Bk Berkelium (247) | **98** Cf Californium (251) | **99** Es Einsteinium (252) | **100** Fm Fermium (257) | **101** Md Mendelevium (258) | **102** No Nobelium (259) | **103** Lr Lawrencium (262) |

Period

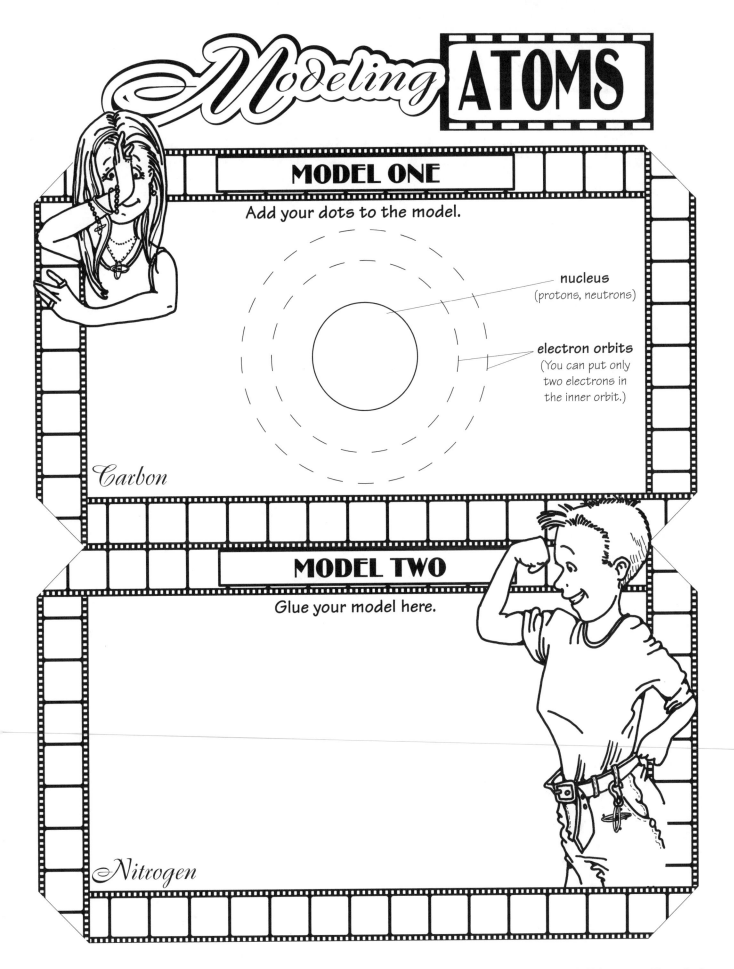

Modeling ATOMS

MODEL ONE

Add your dots to the model.

nucleus
(protons, neutrons)

electron orbits
(You can put only
two electrons in
the inner orbit.)

Carbon

MODEL TWO

Glue your model here.

Nitrogen

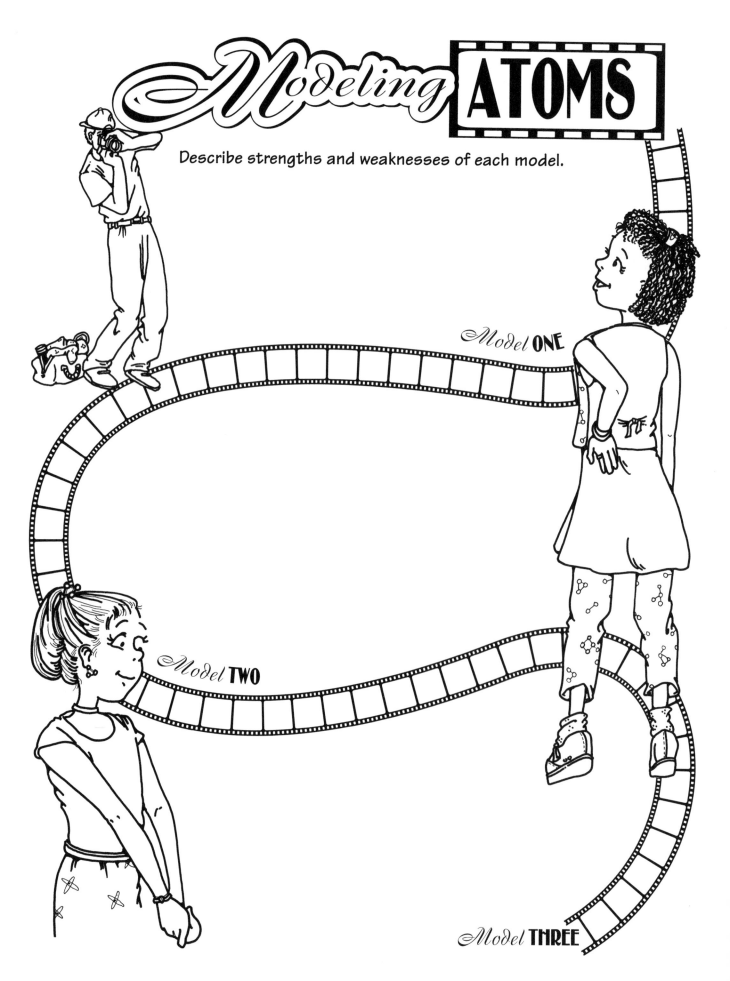

Modeling ATOMS

Describe strengths and weaknesses of each model.

Model ONE

Model TWO

Model THREE

Connecting Learning

1. What are the three different parts of an atom?

2. How did the periodic table help you in this activity?

3. Why do scientists use models of atoms?

4. What do the models show? What do the models not show?

5. What will happen if we learn new things about the structure of atoms?

6. What are you wondering now?

Element Cards

Information cards have been provided as a way to become acquainted with 18 elements. Most of the elements have somewhat familiar names to students, although students may not have any working knowledge of these elements. The cards contain several interesting facts about each one. These facts are only a means to stimulate interest and to perhaps promote further study; they are not intended as content to be memorized.

Information Included on the Cards

- element name in rebus form, as an abbreviation that is used on the periodic table, and in written form along with its pronunciation;
- the element's atomic number and atomic mass;
- a pictorial representation of the element's electrons in their orbital shells;
- the element's state as it appears on Earth at room temperature and under standard pressure;
- some facts—color, reactivity, etc.—about the element;
- who discovered it, when it was discovered, and the origin of its name.

Assembly

Copy the cards on cardstock and cut them apart. Fold each card in half and glue or tape so that the print is on the outside of the card.

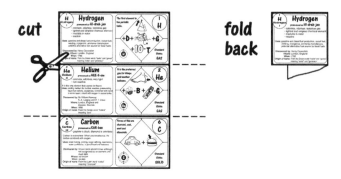

Some Things To Do

Although the use of these cards is often best left to the imagination of students, here are some things that they might do:

- arrange them in rows by the number of orbital shells
- order them by the dates of discovery
- classify them by their standard states
- play "guess me" by giving clues about an element
- connect card to world map with string to identify location of discovery.

Hydrogen

1
H
Hydrogen
1

pronounced as **HI-dreh-jen**

- colorless, odorless, tasteless gas
- lightest and simplest chemical element
- insoluble in water
- reactive

Uses: gasoline and diesel fuel production, rocket fuel, welding, cryogenics, ammonia manufacture, potential alternative fuel source for fossil fuels.

Discovered by: Henry Cavendish
Where: London, England
When: 1766
Origin of name: From the Greek words "hydro" and "genes" meaning "water" and "generator"

The first element in the periodic table.

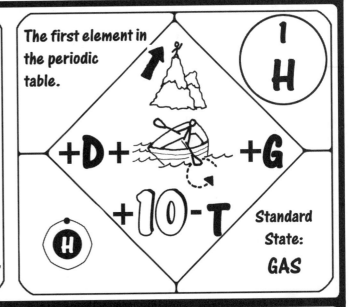

Standard State:
GAS

Helium

2
He
Helium
4

pronounced as **HEE-li-em**

- colorless, odorless, very light
- non-reactive

It is the only element that cannot be frozen.
Uses: cooling medium for nuclear reacters, pressurizing liquid fuel rockets, cryogenics, combined with neon in some lasers, mixed with oxygen in scuba tanks.

Discovered by: Sir William Ramsay
N. A. Langley and P. T. Cleve
Where: London, England and
Uppsala, Sweden
When: 1895
Origin of name: From the Greek word "helios" meaning "sun"

It is the preferred gas for blimps and weather balloons.

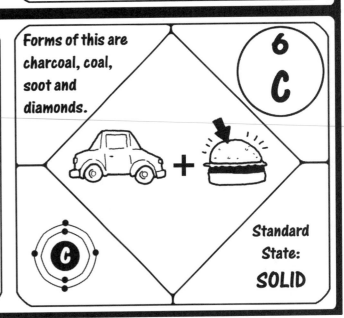

Standard State:
GAS

Carbon

6
C
Carbon
12

pronounced as **KAR-ben**

- graphite is black, diamond is colorless

Carbon is everywhere. When you breathe out, it's carbon combined with oxygen.

Uses: steel making, printing, sugar refining, respirators, water purification, in pencil lead and batteries.

Discovered by: Known since ancient times although not recognised as an element until much later.
Where: not known
When: no data
Origin of name: From the Latin word "carbo" meaning "charcoal"

Forms of this are charcoal, coal, soot and diamonds.

Standard State:
SOLID

Nitrogen

7
N
Nitrogen
14

pronounced as **NYE-treh-gen**
- colorless, odorless, tasteless
- fairly unreactive

Uses: a key component in explosives, rocket fuels, plastics, drugs and dyes and an essential element in fertilizer, and the production of ammonia.

Discovered by: Daniel Rutherford
Where: Scotland
When: 1772
Origin of name: From the Greek words "nitron genes" meaning "nitre" and "forming"

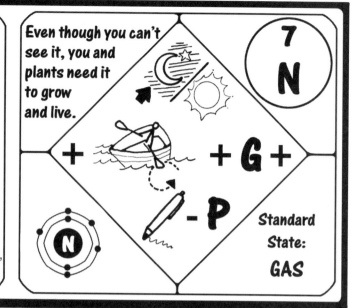

Even though you can't see it, you and plants need it to grow and live.

+ G +

- P

Standard State:
GAS

7
N

Oxygen

8
O
Oxygen
16

pronounced as **OK-si-jen**
- colorless as a gas, liquid is pale blue
- odorless, tasteless
- very reactive

Uses: steel-making, metal-cutting and in medical treatment.

Discovered by: Joseph Priestley, Carl Scheele
Where: England, Sweden
When: 1774
Named by: Antoine Lavoisier in 1777
Origin of name: From the Greek words "oxy genes" meaning "acid" (sharp) and "forming" (acid former)

You breathe it and need it to live. If you breathe too much of it, you could die.

- B

+ - B +

- eral

Standard State:
GAS

8
O

Sodium

11
Na
Sodium
23

pronounced as **SO-di-em**
- very reactive
- oxidizes rapidly when exposed to air
- soft, silvery-white metal

Uses: making drugs, organic compounds, dyes, and in street lights, providing a yellow-orange light.
It is a very light. It's light enough to float on water!

Discovered by: Sir Humphrey Davy
Where: England
When: 1807
Origin of name: From the English word "soda" (the origin of the symbol Na comes from the Latin word "natrium")

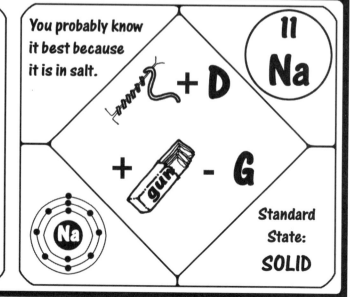

You probably know it best because it is in salt.

+ D

+ gum - G

Standard State:
SOLID

11
Na

Magnesium

12 Mg Magnesium **24**

Magnesium
pronounced as **mag-NEE-zhi-em**

- silvery white, lusterous, soft metal
- reactive
- very light

Uses: if bright, hot flames are needed, such as fireworks, flares, and incendiary weapons.

Discovered by: Sir Humphry Davy
Where: England
When: 1808
Origin of name: From the Greek word "Magnesia," a district of Thessaly, Greece

It was used in the famous milk of magnesia.

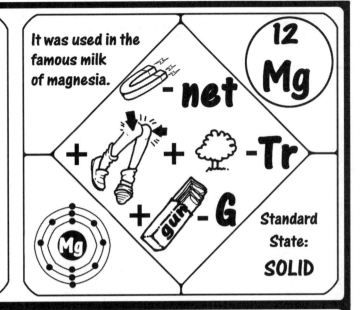

-net
-Tr
-G

12 Mg

Standard State: SOLID

Aluminum

13 Al Aluminum **27**

Aluminum
pronounced as **ah-LOO-men-em**

- silvery-white, relatively soft metal
- very common

Uses: when aluminum is combined with other metals it becomes very strong. It is so strong that engineers use it to build planes and ships.

Discovered by: Hans Christian Oersted
Where: Denmark
When: 1825
Origin of name: From the Latin word "alumen" meaning "alum"

It is used to make cans and foil.

A+
-ig+
-tt +
3,5,2 1,6,4 -ber

13 Al

Standard State: SOLID

Sulfur

16 S Sulfur **32**

Sulfur
pronounced as **SUL-fer**

- non-reactive
- yellowish
- no taste

Uses: brittle, often found as a crystal. It can appear in many different colors, such as orange, brown or red.

Discovered by: Known since ancient times
Where: not known
When: no data
Named by: Antoine Lavoisier in 1777
Origin of name: From the Sanskrit word "sulvere" meaning "sulphur;" also from the Latin word "sulphurium" meaning "sulphur"

It is used to bleach dried fruits and paper goods.

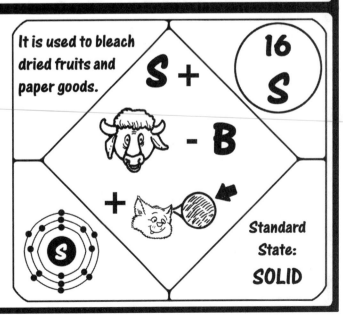

S +
- B
+

16 S

Standard State: SOLID

Chlorine

17
Cl
Chlorine
36

pronounced as **KLOR-een** or as **KLOR-in**

- yellowish green
- disagreeable gas

Uses: production of safe drinking water, production of paper products, dyes, textiles, plastics, medicines, antiseptics, insecticides, foodstuff, solvents, and paints.

Discovered by: Carl William Scheele
Where: Sweden
When: 1774
Named by: Sir Humphry Davy in 1810
Origin of name: From the Greek word "chloros" meaning "pale green"

It is the smell of household bleach or a swimming pool.

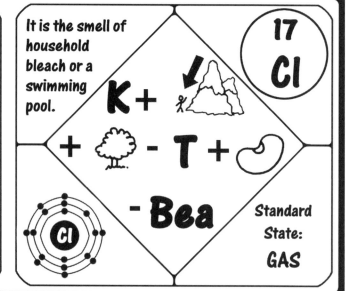

17
Cl

Standard State:

GAS

Potassium

19
K
Potassium
39

pronounced as **pe-TASS-i-em**

- soft, light, waxy, silvery-white metal
- violently reactive

It tarnishes to a grayish color in seconds.

Uses: in match heads, glass, soaps, explosives, baking powder, tanning lotions, and especially in fertilizers.

Discovered by: Sir Humphrey Davy
Where: England
When: 1807
Origin of name: From the English word "potash" (pot ashes) the origin of the symbol K comes from the Latin word "kalium"

It is used as a salt substitute.

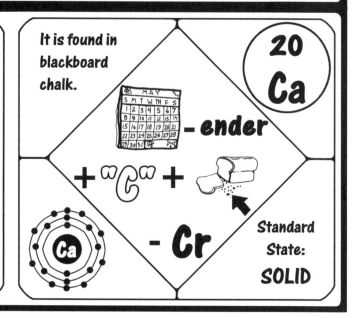

19
K

Standard State:

SOLID

Calcium

20
Ca
Calium
40

pronounced as **KAL-si-em**

- silvery-white metal
- very reactive

Uses: to make white paint, cleaning powder, toothpaste, and stomach antacids.

It forms part of cell walls, teeth, and bones. It is also important for blood clotting.

Discovered by: Sir Humphrey Davy
Where: England
When: 1808
Origin of name: From the Latin word "calx" meaning "lime"

It is found in blackboard chalk.

20
Ca

Standard State:

SOLID

22 Ti Titanium 48

Titanium

pronounced as **tie-TAY-ni-em**

- hard, lusterous, silvery, dark-gray metal
- stable

Uses: airframes, engines, as a component of joint replacement parts, a pigment to create white paint, a yellow food additive, create an artificial gemstone, and in sun screen lotions.

Discovered by: Reverend William Gregor
Where: England
When: 1791
Origin of name: Named after the "Titans," (the sons of the Earth goddess in Greek mythology)

It is as strong as steel, but much lighter.

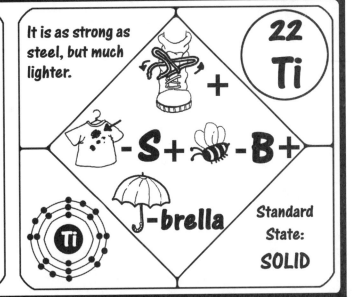

Standard State: **SOLID**

26 Fe Iron 56

Iron

pronounced as **EYE-ern**

- lustrous, silvery, soft, metal
- strongly reactive

Meteorites are one of the few places you will find a crystalline form of Iron.

Uses: to make paper clips, skyscrapers and everything in between.

Discovered by: Known since ancient times
Where: not known
When: no data
Origin of name: From the Anglo-Saxon word "iron" the origin of the symbol Fe comes from the Latin word "ferrum" meaning "iron"

It is the cheapest, most abundant, useful, and important of all metals.

Standard State: **SOLID**

27 Co Cobalt 59

Cobalt

pronounced as **KO-bolt**

- hard, brittle, lustrous, silvery-blue metal
- stable

Uses: making powerful permanent magnets and stainless steels.

Discovered by: Georg Brandt
Where: Sweden
When: 1735
Origin of name: From the German word "kobald" meaning "goblin" or evil spirit

Used to make the blue colors in porcelain, glass, pottery, tiles, and enamels.

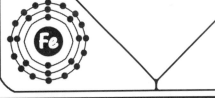

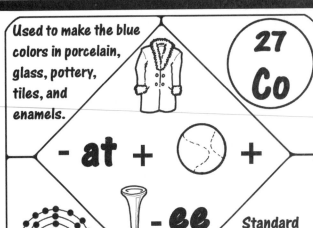

Standard State: **SOLID**

28
Ni
Nickel
59

Nickel

pronounced as **NIK-'l**

- lustrous, hard, malleable, silvery-white metal
- stable

Uses: to provide a protective coating for other metals, and for armor plates and burglar-proof vaults.

Discovered by: Baron Axel Fredrik Cronstedt
Where: Sweden
When: 1751
Origin of name: From the German word "kupfernickel" meaning Devil's copper or St Nicholas's (Old Nick's) copper

It is used to make the five-cent coin.

28
Ni

N + 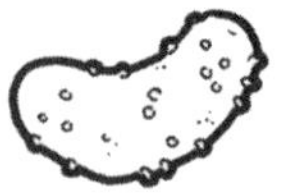 **- P**

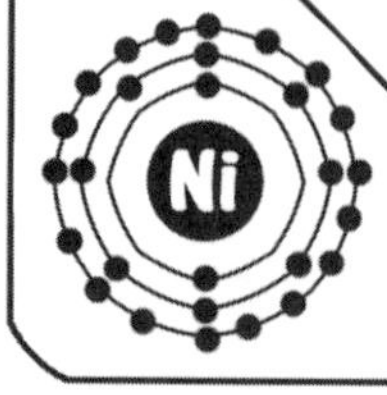

Standard State: **SOLID**

29
Cu
Copper
64

Copper

pronounced as **KOP-er**

- malleable, reddish-brown, bright metallic luster
- copper that has oxidized is bluish
- stable

Uses: around 6000BC it was being used in pottery in North Africa. The electrical industry is one of the largest users of copper.

Discovered by: Known since ancient times
Where: not known
When: no data
Origin of name: From the Latin word "cuprum" meaning the island of "Cyprus"

The penny is plated with this.

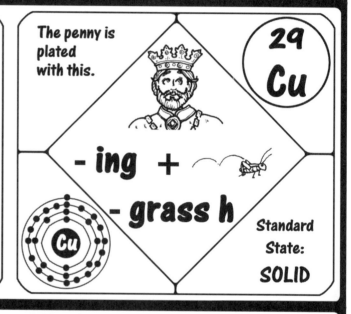

29
Cu

- ing + 

- grass h

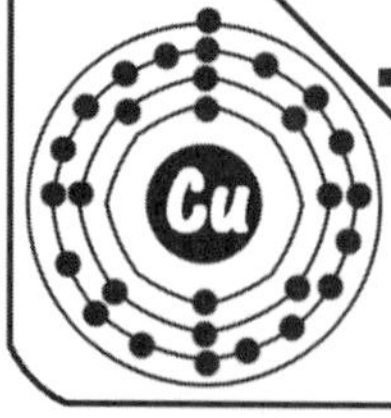

Standard State: **SOLID**

36
Kr
Krypton
84

Krypton

pronounced as **KRIP-ton**

- colorless, odorless, heavy gas
- unreactive

Uses: it is used in some photographic flash lamps for high-speed photography. It is one of the filling gases for fluorescent lights.

Discovered by: Sir William Ramsay and Morris W. Travers
Where: Great Britan
When: 1898
Origin of name: From the Greek word "kryptos" meaning "hidden"

The similarity to Superman's strength-sapping nemesis is just coincidence.

36
Kr

K + 

+ 

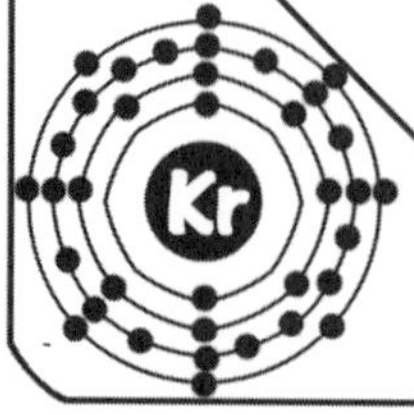

Standard State: **GAS**

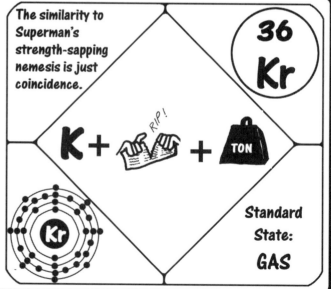

The AIMS Program

AIMS is the acronym for "Activities Integrating Mathematics and Science." Such integration enriches learning and makes it meaningful and holistic. AIMS began as a project of Fresno Pacific University to integrate the study of mathematics and science in grades K-9, but has since expanded to include language arts, social studies, and other disciplines.

AIMS is a continuing program of the non-profit AIMS Education Foundation. It had its inception in a National Science Foundation funded program whose purpose was to explore the effectiveness of integrating mathematics and science. The project directors in cooperation with 80 elementary classroom teachers devoted two years to a thorough field-testing of the results and implications of integration.

The approach met with such positive results that the decision was made to launch a program to create instructional materials incorporating this concept. Despite the fact that thoughtful educators have long recommended an integrative approach, very little appropriate material was available in 1981 when the project began. A series of writing projects have ensued and today the AIMS Education Foundation is committed to continue the creation of new integrated activities on a permanent basis.

The AIMS program is funded through the sale of this developing series of books and proceeds from the Foundation's endowment. All net income from program and products flows into a trust fund administered by the AIMS Education Foundation. Use of these funds is restricted to support of research, development, and publication of new materials. Writers donate all their rights to the Foundation to support its on-going program. No royalties are paid to the writers.

The rationale for integration lies in the fact that science, mathematics, language arts, social studies, etc., are integrally interwoven in the real world from which it follows that they should be similarly treated in the classroom where we are preparing students to live in that world. Teachers who use the AIMS program give enthusiastic endorsement to the effectiveness of this approach.

Science encompasses the art of questioning, investigating, hypothesizing, discovering, and communicating. Mathematics is a language that provides clarity, objectivity, and understanding. The language arts provide us powerful tools of communication. Many of the major contemporary societal issues stem from advancements in science and must be studied in the context of the social sciences. Therefore, it is timely that all of us take seriously a more holistic mode of educating our students. This goal motivates all who are associated with the AIMS Program. We invite you to join us in this effort.

Meaningful integration of knowledge is a major recommendation coming from the nation's professional science and mathematics associations. The American Association for the Advancement of Science in *Science for All Americans* strongly recommends the integration of mathematics, science, and technology. The National Council of Teachers of Mathematics places strong emphasis on applications of mathematics such as are found in science investigations. AIMS is fully aligned with these recommendations.

Extensive field testing of AIMS investigations confirms these beneficial results.

1. Mathematics becomes more meaningful, hence more useful, when it is applied to situations that interest students.

2. The extent to which science is studied and understood is increased, with a significant economy of time, when mathematics and science are integrated.

3. There is improved quality of learning and retention, supporting the thesis that learning which is meaningful and relevant is more effective.

4. Motivation and involvement are increased dramatically as students investigate real-world situations and participate actively in the process.

We invite you to become part of this classroom teacher movement by using an integrated approach to learning and sharing any suggestions you may have. The AIMS Program welcomes you!

AIMS Education Foundation Programs

A Day with AIMS®

Intensive one-day workshops are offered to introduce educators to the philosophy and rationale of AIMS. Participants will discuss the methodology of AIMS and the strategies by which AIMS principles may be incorporated into curriculum. Each participant will take part in a variety of hands-on AIMS investigations to gain an understanding of such aspects as the scientific/mathematical content, classroom management, and connections with other curricular areas. *A Day with AIMS®* workshops may be offered anywhere in the United States. Necessary supplies and take-home materials are usually included in the enrollment fee.

A Week with AIMS®

Throughout the nation, AIMS offers many one-week workshops each year, usually in the summer. Each workshop lasts five days and includes at least 30 hours of AIMS hands-on instruction. Participants are grouped according to the grade level(s) in which they are interested. Instructors are members of the AIMS Instructional Leadership Network. Supplies for the activities and a generous supply of take-home materials are included in the enrollment fee. Sites are selected on the basis of applications submitted by educational organizations. If chosen to host a workshop, the host agency agrees to provide specified facilities and cooperate in the promotion of the workshop. The AIMS Education Foundation supplies workshop materials as well as the travel, housing, and meals for instructors.

AIMS One-Week Perspectives Workshops

Each summer, Fresno Pacific University offers AIMS one-week workshops on its campus in Fresno, California. AIMS Program Directors and highly qualified members of the AIMS National Leadership Network serve as instructors.

The AIMS Instructional Leadership Program

This is an AIMS staff-development program seeking to prepare facilitators for leadership roles in science/math education in their home districts or regions. Upon successful completion of the program, trained facilitators may become members of the AIMS Instructional Leadership Network, qualified to conduct AIMS workshops, teach AIMS in-service courses for college credit, and serve as AIMS consultants. Intensive training is provided in mathematics, science, process and thinking skills, workshop management, and other relevant topics.

College Credit and Grants

Those who participate in workshops may often qualify for college credit. If the workshop takes place on the campus of Fresno Pacific University, that institution may grant appropriate credit. If the workshop takes place off-campus, arrangements can sometimes be made for credit to be granted by another institution. In addition, the applicant's home school district is often willing to grant in-service or professional-development credit. Many educators who participate in AIMS workshops are recipients of various types of educational grants, either local or national. Nationally known foundations and funding agencies have long recognized the value of AIMS mathematics and science workshops to educators. The AIMS Education Foundation encourages educators interested in attending or hosting workshops to explore the possibilities suggested above. Although the Foundation strongly supports such interest, it reminds applicants that they have the primary responsibility for fulfilling *current* requirements.

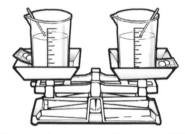

For current information regarding the programs described above, please complete the following:

Information Request

Please send current information on the items checked:

____ *Basic Information Packet* on AIMS materials   ____ *A Week with AIMS®* workshops
____ *AIMS Instructional Leadership Program*   ____ Hosting information for *A Day with AIMS®* workshops
____ *AIMS One-Week Perspectives* workshops   ____ Hosting information for *A Week with AIMS®* workshops

Name _____ Phone _____

Address _____
       Street   City   State   Zip
